# NOVEL DAIRY PROCESSING TECHNOLOGIES

## TECHNOLOGIES

Techniques, Management, and Energy Conservation

*Innovations in Agricultural and Biological Engineering*

# NOVEL DAIRY PROCESSING TECHNOLOGIES

Techniques, Management, and Energy Conservation

*Edited by*
**Megh R. Goyal, PhD**
**Anit Kumar, MTech**
**Anil K. Gupta, PhD**

APPLE
ACADEMIC
PRESS

Apple Academic Press Inc.
3333 Mistwell Crescent
Oakville, ON L6L 0A2 Canada

Apple Academic Press Inc.
9 Spinnaker Way
Waretown, NJ 08758 USA

© 2018 by Apple Academic Press, Inc.

First issued in paperback 2021

*Exclusive worldwide distribution by CRC Press, a member of Taylor & Francis Group*
No claim to original U.S. Government works

ISBN 13: 978-1-77-463635-0 (pbk)
ISBN 13: 978-1-77-188612-3 (hbk)

---

### Library and Archives Canada Cataloguing in Publication

---

Novel dairy processing technologies : techniques, management, and energy conservation / edited by Megh R. Goyal, PhD, Anit Kumar, MTech, Anil K. Gupta, PhD.

(Innovations in agricultural and biological engineering)
Includes bibliographical references and index.
Issued in print and electronic formats.
ISBN 978-1-77188-612-3 (hardcover).--ISBN 978-1-315-16712-1 (PDF)

1. Dairy products industry--Technological innovations. 2. Dairy processing. 3. Milk--Pasteurization. I. Goyal, Megh Raj, editor II. Kumar, Anit, editor III. Gupta, Anil K., 1966- editor IV. Series: Innovations in agricultural and biological engineering

| SF250.5.N68 2017 | 637 | C2017-907441-5 | C2017-907442-3 |

---

### Library of Congress Cataloging-in-Publication Data

---

Names: Goyal, Megh Raj, editor. | Kumar, Anit, editor. | Gupta, Anil K. (Anil Kumar), 1966- editor.
Title: Novel dairy processing technologies : techniques, management, and energy conservation / editors: Megh R. Goyal, Anit Kumar, Anil K. Gupta.
Description: Waretown, NJ : Apple Academic Press, 2017. | Series: Innovations in agricultural & biological engineering ; volume 17 | Includes bibliographical references and index.
Identifiers: LCCN 2017055545 (print) | LCCN 2017056542 (ebook) | ISBN 9781315167121 (ebook) | ISBN 9781771886123 (hardcover : alk. paper)
Subjects: LCSH: Dairy processing. | Dairying.
Classification: LCC SF250.5 (ebook) | LCC SF250.5 .N684 2017 (print) | DDC 338.4/7637--dc23
LC record available at https://lccn.loc.gov/2017055545

---

Apple Academic Press also publishes its books in a variety of electronic formats. Some content that appears in print may not be available in electronic format. For information about Apple Academic Press products, visit our website at **www.appleacademicpress.com** and the CRC Press website at **www.crcpress.com**

# CONTENTS

## PART I: Emerging Processing Technologies in the Dairy Industry ... 1

# LIST OF CONTRIBUTORS

**Adarsh M. Kalla**
Department of Dairy Engineering, Dairy Science College, Karnataka Veterinary Animal and Fisheries Sciences University, Mahagoan Cross, Kalaburagi 585316, Karnataka India. E-mail: adarshkalla002@gmail.com

**Ajay Kumar Kashyap**
Department of Agricultural and Environmental Science, National Institute of Food Technology Entrepreneurship and Management (NIFTEM) Kundli, Haryana, India

**Amit Juneja**
National Institute of Food Technology Entrepreneurship and Management, Kundli, Sonepat 131028 Haryana, India

**Angadi Viswanatha**
Department of Food Microbiology, Agriculture College Hassan, University of Agricultural Sciences (UAS), Bangalore 573225, India. E-mail: angadigm@gmail.com

**Anil K. Gupta**
Department of Agricultural Economics, College of Agriculture, CCS Haryana Agricultural University, Hisar 125004, Haryana, India. E-mail: doc@agdentalclinic.com gotogupta@hotmail.com

**Anit Kumar**
Department of Food Science and Technology, National Institute of Food Technology Entrepreneurship and Management, Plot No. 97, Sector 56, HSIIDC Industrial Estate, Kundli, Sonipat 131028, Haryana, India. E-mail: aks.kumar6@gmail.com

**Asaad Rehman Saeed Al-Hilphy**
Department of Food Science, College of Agriculture, University of Basrah, Basrah City, Iraq. E-mail: aalhilphy@yahoo.co.uk

**Ashok K. Agrawal**
Department of Dairy Engineering, College of Dairy Science and Food Technology, Chhattisgarh Kamdhenu Vishwavidyalaya (CGKV), Raipur 492012 India, E-mail: akagrawal.raipur@gmail.com

**Ashutosh Upadhyay**
National Institute of Food Tech. Entrepreneurship & Management (NIFTEM), Plot No. 97, Sector 56, HSIIDC Industrial Estate Kundli, Sonipat 131028, Haryana India. E-mail: ashutosh.niftem@gmail.com

**Bhavesh B. Chavhan**
Department of Dairy Engineering, College of Dairy Technology Maharashtra Animal and Fisheries Sciences University, Udgir, Nagpur 413517, India. E-mail: bhaveshchavhan@gmail.com; bhaveshchavhan90@gmail.com

**C. Ram Kumar**
Department of Food Science and Technology, College of Agriculture, Hassan, University of Agricultural Sciences, Bangalore 573225, India

**C. Sahu**
Department of Dairy Engineering, College of Dairy Science and Food Technology, Chhattisgarh Kamdhenu Vishwa Vidyalaya (CGKV), Raipur 492012, India

**Chandan Solanki**
ICAR-Central Institute of Post-Harvest Engineering and Technology, Ludhiana 141001, Punjab, India. E-mail: chandan4uu12@gmail.com

**D. C. Saxena**
Sant Longowal Institute of Engineering & Technology, Longowal, 148106, Sangrur Punjab, India. E-mail: dcsaxena@yahoo.com

**Devaraju Rajanna**
Department of Dairy Engineering, Karnataka Veterinary Animal and Fisheries Sciences University, Dairy Science College, Mahagoan Cross, Kalaburagi 585316, Karnataka, India

**Durga Shankar Bunkar**
Centre of Food Science and Technology, Institute of Agricultural Sciences, Banaras Hindu University, Varanasi 221005, India

**G. Bhadania**
Department of Dairy Engineering, Sheth M. C. College of Dairy Science, Anand Agricultural University, Anand 388110, India. E-mail: bhadania@gmail.com

**Ghassan F. Mohsin**
Vocational School, Maysan City, Iraq. E-mail: ghassanmohsin@ymail.com

**Gurdeep Rattu**
Department of Basic and Applied Science, National Institute of Food Technology Entrepreneurship and Management (NIFTEM) Kundli, Haryana, India

**Haider I. Ali**
Department of Food Science, College of Agriculture, University of Basrah, Basra City, Iraq.

**Hitesh Patel**
Dairy Chemistry Division, National Dairy Research Institute, Karnal 132001, Haryana, India

**Jayeeta Mitra**
Department of Agricultural and Food Engineering, Indian Institute of Technology Kharagpur, Kharagpur 721302, India. E-mail: jayeeta.mitra@agfe.iitkgp.ernet.in; jayeeta12@gmail.com

**Kumaresh Halder**
National Institute of Food Technology Entrepreneurship and Management, Kundli, Sonepat 131028 Haryana, India. E-mail: kumar.halder@gmail.com

**Megh R. Goyal**
Retired Faculty in Agricultural and Biomedical Engineering from General Engineering Department, University of Puerto Rico – Mayaguez Campus; and Senior Technical Editor-in-Chief in Agriculture Sciences and Biological Engineering, Apple Academic Press Inc., USA.
E-mail: goyalmegh@gmail.com

**Mohammed Nayeem**
Department of Food Science and Technology, National Institute of Food Technology Entrepreneurship and Management (NIFTEM) Kundli, Haryana, India.
E-mail: nayeem.amaan@gmail.com

**Nishant Kumar**
Department of Agricultural and Environmental Science, National Institute of Food Technology Entrepreneurship and Management (NIFTEM), Kundli, Haryana, India

**P. P. Sutar**
Department of Food Process Engineering, National Institute of Technology, Rourkela 769008, Odisha, India. E-mail: sutarp@nitrkl.ac.in; paragsutar@gmail.com

**Prabhat Kumar Nema**
Department of Food Engineering, National Institute of Food Technology Entrepreneurship and Management, Plot No. 97, Sector 56, HSIIDC Industrial Estate, Kundli, Sonipat 131028, Haryana

**Pramod Kumar**
Dairy Chemistry Division, National Dairy Research Institute (NDRI), Karnal 132001, Haryana, India. E-mail: param.bhu@gmail.com

**Rachna Sehrawat**
Department of Food Engineering, National Institute of Food Technology Entrepreneurship and Management, Plot No. 97, Sector 56, HSIIDC Industrial Estate, Kundli, Sonipat 131028, Haryana, India. E-mail: sehrawatrachna@gmail.com

**Raman Seth**
Dairy Chemistry Division, National Dairy Research Institute, Karnal 132001, Haryana, India

**Rohant Dhaka**
Department of Food Science and Technology, National Institute of Food Technology Entrepreneurship and Management (NIFTEM) Kundli, Haryana, India

**S. S. Roopa**
Britannia Industries Ltd., Plot # 23, Bidadi Industrial Area, Bidadi Hobli, Ramanagara 562109 Karnataka, India

**S. S. Shirkole**
Department of Food Process Engineering, National Institute of Technology, Rourkela 769008, Odisha, India

**Santosh S. Chopde**
Department of Dairy Engineering, College of Dairy Technology at Udgir, COVAS Campus, Kavalkhed Road, Udgir 413517 (MS), India

**Shilpi Singh**
Food Process Engineering, Department of Agricultural and Food Engineering, Indian Institute of Technology, Kharagpur 721302, India. E-mail: shilpianjel@gmail.com

**Subrota Hati**
SMC College of Dairy Science, Anand Agricultural University (AAU), Anand 388001, Gujarat, India

**Vijendra Mishra**
Department of Basic and Applied Sciences, National Institute of Food Tech. Entrepreneurship & Management (NIFTEM), Plot No. 97, Sector 56, HSIIDC Industrial Estate, Kundli, Sonipat 131028, Haryana, India

# LIST OF ABBREVIATIONS

| | |
|---|---|
| Ab | antibody |
| ALP | associated alkaline phosphatase |
| ASHARE | American Society of Heating, Air Conditioning and Refrigerating Engineers |
| ATP | adenosine tri phosphate |
| AU | American unit |
| BEE | Bureau of Energy Efficiency |
| BSA | bovine serum albumin |
| bST | bovine somatotropin |
| CCL | critical control limit |
| CCP | critical control point |
| CFU | colony forming units |
| CGKV | Chhattisgarh Kamdhenu Vishwavidyalaya |
| CGR | compound growth rate |
| CIP | clean-in-place |
| CMA | cow's milk allergy |
| $CO_2$ | carbon dioxide |
| COD | chemically oxygen demand |
| COP | coefficient of performance |
| CRI | color rendition index |
| CRIE | crossed radio-immuno electrophoresis |
| CV | coefficient of variation |
| D | diffusion coefficient of water from homogeneous material during osmotic dehydration |
| $D_1$ | diffusion coefficients of water from the core of the material to the dehydration front |
| $D_2$ | diffusion coefficients of water from the core of the material to across the front |
| $D_3$ | diffusion coefficients of water from the core of the material through the osmotically treated material into the osmotic solution |
| DM | dry matter |
| EpiPen | epinephrine |

| ESL | extended shelf life |
| ETP | effluent treatment plant |
| Fc | foot candles |
| FFA | free fatty acids |
| FSMS | food safety management system |
| FSSAI | Food Safety and Standards Authority of India |
| GC | gas chromatography |
| GC–MS | gas chromatography and mass spectroscopy |
| GHG | greenhouse gas |
| GMP | good manufacturing practices |
| HACCP | hazard analysis and critical control point |
| HDDCF | Haryana Dairy Development Cooperative Federation Ltd. |
| HP | high pressure |
| HPH | high pressure homogenization |
| HPP | high pressure processing |
| HPS | high pressure sodium |
| HTST | high temperature short time |
| HX | heat exchanger |
| ICAR | Indian Council of Agricultural Research |
| IDF | International Dairy Federation |
| Ig | immunoglobin |
| IgE | immunoglobulin E |
| IGF | insulin-like growth factor |
| INR | Indian rupees |
| ISO | International Organization for Standards |
| IU | international unit |
| KDa | kilodalton |
| kV | kilovolt |
| L/d | liters per day |
| LAB | lactic acid bacteria |
| LDL | long-day lighting |
| Lf | lactoferrin |
| LP | low pressure |
| LTLT | low temperature long time |
| $M/M_O$ | relative moisture content |
| MH | metal halide |
| MSNF | milk solids not fat |
| MSU | Michigan State University |

| | |
|---|---|
| NACICMR | Nutritional Advisory Committee of the Indian Council of Medical Research |
| NCA | National Commission on Agriculture |
| NDRI | National Dairy Research Institute |
| NGOs | non-government agencies |
| NIFTEM | National Institute of Food Technology Entrepreneurship and Management |
| NIR | near infrared |
| NLT | not less than |
| NMT | not more than |
| OFP | operation flood program |
| OPRP | operational pre-requisite program |
| PEF | pulse electric field |
| PFA | prevention of food adulteration |
| PRL | hormone prolactin |
| PRP | pre-requisite program |
| RHT | reduced heat treatment |
| RLU | relative light units |
| SG | specific gravity |
| SNF | solids not fat |
| SPC | system performance coefficient |
| SSOPs | sanitary standard operating procedures |
| TBA | thiobarbituric acid |
| TES | thermal energy storage system |
| TNDC | Tamil Nadu Dairy Development Corporation |
| TR | ton of refrigeration |
| UASB | up-flow anaerobic sludge blanket reactor |
| UASI | ultrasonication-assisted spray ionization |
| UHT | ultra-high temperature |
| USFDA | United States Food and Drug Administration |
| VCRS | vapor compressor refrigeration system |
| $Z_P$ | cell disintegration index |
| α-Lac | alpha-lactalbumin |
| β-Lg | beta-lactoglobulin |
| $\Delta x$ | thickness of moving dehydration front |

# PREFACE 1

Milk is nutritious, appetizing, and nature's perfect food; it is recommended by nutritionists for the development of a sound body and is consumed by all sectors of people. Demand for milk is greater in developed countries as compared to developing countries, but the gap has narrowed due to an increase in urbanization, population, and consumption. India ranks first in milk production, accounting for 18.5% of world production, achieving an annual output of 146.3 million tons during 2014–2015 as compared to 137.69 million tons during 2013–2014, recording a growth of 6.26%, whereas the Food and Agriculture Organization (FAO) has reported a 3.1% increase in world milk production from 765 million tons in 2013 to 789 million tons in 2014. Pasteurized milk and ultra-high temperature (UHT) milk market share in 2014 has increased to 83.7 and 6.9% (by volume), respectively, compared to 81.9 and 5.2% in 2013.

A *milk product* is produced from the milk of mammals. Dairy products are usually high energy-yielding food products. A production plant for the processing of milk is called a dairy or a dairy factory. Apart from breastfed infants, the human consumption of dairy products is sourced primarily from the milk of cows, water buffaloes, goats, sheep, yaks, horses, camels, domestic buffaloes, and other mammals. Dairy products are commonly found throughout the world.

I recall my childhood. I was breastfed by mother until I was 8 years old. It calmed me down from my nervousness and hypertension and imparted to me security in the lap of my mother. Until I was in fifth grade, I and my paternal family never tasted dairy milk except a few drops in an Indian tea. In sixth grade, once a day, I and my classmates were asked to sit on a $1 \times 50$ m$^2$ mat and were served one 16-oz glass of milk (that was prepared from milk powder of American origin) by American Peace Corps volunteers. I enjoyed not only drinking the milk but also the dedication of these volunteers. What a gesture to the undernourished students in developing countries!

At the 49th annual meeting of the Indian Society of Agricultural Engineers at the Punjab Agricultural University (PAU) during February 22–25 of 2015, a group of ABEs and FEs convinced me that there is a dire need to publish book volumes on focus areas of agricultural and biological engineering (ABE). This is how the idea was born on the new book series titled Innovations in Agricultural & Biological Engineering. This book volume 17, *Novel Dairy Processing Technologies: Techniques, Management and Energy Conservation*, under this book series contributes to the ocean of knowledge on dairy engineering.

The contributions by the cooperating authors to this book volume have been most valuable in the compilation. Their names are mentioned in each chapter and in the list of contributors. This book would not have been written without the valuable cooperation of these investigators, many of whom are renowned scientists who have worked in the field of dairy engineering throughout their professional careers.

I am glad to introduce Anit Kumar, who is at present a PhD research scholar in the Department of Food Science and Technology at the National Institute of Food Technology Entrepreneurship and Management, Kundli, Haryana, India. His propensity for learning, inquiry and discovery of new facts has inspired him to opt for independent research as a career in dairy science.

Dr. Anil K. Gutpa contributed his editorial skills to this book volume. Without their support and leadership qualities as co-editors of this book volume and the extraordinary work on dairy engineering applications, readers will not have this quality publication.

I would like to thank the editorial staff, Sandy Jones Sickels, Vice President, and Ashish Kumar, Publisher and President, at Apple Academic Press, Inc., for making every effort to publish the book when food and water issues are major issues worldwide. Special thanks are also due to the AAP Production Staff.

I request that readers to offer constructive suggestions that may help to improve the next edition.

I express my deep admiration to my family for their understanding and collaboration during the preparation of this book volume. As an educator, there is a piece of advice to one and all in the world: "*Permit that our almighty God, our Creator, provider of all and excellent Teacher, feed our life with Healthy Milk and Milk Products and His Grace—and Get married to your profession*"

—**Megh R. Goyal, PhD, PE**
Senior Editor-in-Chief

# PREFACE 2

Dairy science has been an attractive discipline for more than 100 years. Milk and milk products are important diets of people throughout the world. The importance of milk as a source of multi-nutritional components is a well-established fact. It is also being used as a raw material by several dairy processing industries. At present, several books are available on the market written on dairy processes, dairy products, hurdles and management in dairy industries but they all seem to be scattered. This book is intended to cover the most important topics, such as processing, products, hurdles, management and energy conservation.

The book consists of four parts. **Part I** is devoted to the application of novel processing technologies in the dairy industry, where four chapters have been included: Technology of ohmic heating for pasteurization of milk, Advances in microwave-assisted processing of milk, Hybrid technology for pasteurization of milk, and Advancement in the ghee making process. **Part II** is "Novel Drying Techniques in the Dairy Industry," with two chapters on classification of dried milk products, and osmotic dehydration: principles and applications in the dairy industry. **Part III** focuses on management systems and hurdles in the dairy industry and includes three chapters on Food safety and management system (FSMS): applications in the dairy industry, Economic analysis of milk marketing: case study in Haryana, and dairy foods: allergy and intolerance. **Part IV** covers "Energy Conservation: Opportunities in the Dairy Industry," with three chapters that focus on Day lighting in dairy farms, Refrigeration principles and applications in the dairy industry, and Energy audits and its potential as a tool for energy conservation in the dairy industry.

We hope that the book will provide clear concepts about novel processing in the dairy sector. It also aims to answer questions related to hurdles, management, and energy conservation in the dairy sector. The book has been particularly written for food science students, researchers, academics, and professionals in the food industry.

I am highly thankful to all the authors of individual chapters for their sincere efforts in writing the assigned book chapter. I also owe my gratitude to Professor Megh R. Goyal for his support and arduous job to bring the book into its final form. I also acknowledge the help from Apple Academic Press Inc. for providing the opportunity to publish this book.

I sincerely acknowledge the support received from Dr. Ajit Kumar (Vice Chancellor, NIFTEM, India); my research guide Dr. Ashutosh Upadhyay (Associate Professor, FST Department, NIFTEM, India); and Co-research guide Dr. Vijendra Mishra (Associate Professor, BAS Department, NIFTEM, India); Dr. Rupesh S. Chavan (Senior Executive, Mother Dairy Fruit and Vegetable Pvt. Ltd., Department of Quality Assurance, Gujarat, India); Dr. P.K. Nema (Associate Professor, FE Department, NIFTEM, India); and Dr. D.C. Saxena (Professor, Food Engineering Department, SLIET, India).

I thank my grandparents, my parents Pramila Singh and Ashok K. Singh, elder brother Amit K. Singh and his wife Poonam Srivastava, my younger brother Anish K. Singh, and my colleagues Rachna Sehrawat, Onkar A. Babar, and Sandeep K. Gaikwad at the National Institute of Food Technology Entrepreneurship and Management, Haryana, India, for their constant love and affection.

—**Anit Kumar**
Co-Editor

# FOREWORD 1 BY D. C. SAXENA

This book provides an insight into innovations and the latest research and policies being adopted in the dairy sector. This sector has immense growing opportunities and challenges for farmers, entrepreneurs, and industrialists as well as for academicians and researchers. The editors have put great effort to present the scientific research on drying of dairy products, management practices, advanced processing technology and hurdles in the sector. The most important highlight is the *"Energy Conservation: Opportunities in the Dairy Industry."* It is the need of the hour to focus on technologies and innovations to reduce the requirements of energy and its utilization in the best possible way. I wish for the success of the book and hope it will be a good source of literature for people who are involved in dairy industry growth and making it stronger day by day and help them to overcome the challenges.

**D. C. Saxena, PhD**
Professor (Food Process Engineering), Former Head of
the Department Food Engineering & Technology, Former
Dean (Planning & Development) Sant Longowal Institute
of Engineering & Technology Longowal, 148106.
District Sangrur (Punjab), India
E-mail: dcsaxena@yahoo.com

# FOREWORD 2 BY ASHUTOSH UPADHYAY

Milk is valued as one of the most complicated food systems, and therefore, extensive studies have been performed on a range of topics including techniques, management and energy conservation in the dairy industry. The dairy industry has been a chief contributor to the manufacturing capacity in many countries. In addition to pasteurized milk, whey protein concentrates and simple curd to specialty probiotic dairy drinks, there are many more new milk-based products in the market, and this is likely to grow further with innovative products ideas tried in the dairy sector.

There are several books on dairy processing and technologies. However, this book, *Novel Dairy Processing Technologies: Techniques, Management and Energy Conservation*, is not just to stand with the crowd as it has a foremost distinction by dealing with various novel techniques of food processing that have emerged during the past five decades. This also makes the book equally useful for a food technologist. Although the book is not a text book, it would be highly useful for university-level education in food/dairy science and technology.

It is important to mention that this book covers novel processing technologies in the dairy sector, such as: Ohmic heating, microwave-assisted processing, hybrid technology, and osmotic dehydration. Products like Kulfi, which is an Indian frozen speciality, has found a unique place in the book. Probably looking into the global importance of food safety, the editors have touched on all relevant topics, such as food management systems, hurdles in the dairy industry, allergens, etc. Energy conservation and energy audits are aptly covered in the book.

The book is a nice bouquet of various timely topics that have relevance and significance for a modern food or dairy technologist. I am also pleased to notice that most of the authors are energetic youths, emerging on the horizon of food and dairy research. I appreciate the publisher and editors for this endeavor.

**Ashutosh Upadhyay, PhD**
Associate Professor (Food Science and Technology)
National Institute of Food Tech. Entrepreneurship
& Management (NIFTEM), Plot No. 97, Sector 56,
HSIIDC Industrial Estate Kundli, District Sonipat,
Haryana – 131028, India Mobile: +91-9034022694
E-mail: ashutosh.niftem@gmail.com

# FOREWORD 3 BY VIJENDER MISHRA

Milk is nature's perfect food; it only lacks iron, copper, and vitamin C, and it is highly recommended by nutritionists for building a healthy body. New technologies have emerged in processing milk. To enlighten the reader, the processing of milk by novel techniques, solutions to different hurdles, management, and conservation of energy in the dairy sector have been emphasized in the book. This book is divided into four parts: Part I: Application of Novel Processing Technologies in the Dairy Industry; Part II: Novel Drying Techniques in the Dairy Industry; Part III: Management Systems and Hurdles in the Dairy Industry; Part IV: Energy Conservation: Opportunities in the Dairy Industry.

It is nearly impossible to have milk with no microorganisms, even with the most ideal sanitation condition of animals or handlers. It has been found that the thermal treatment of milk leads to the preservation of physiological properties of milk. The most common thermal treatments used by the dairy industry are pasteurization and sterilization. This book focuses on the technology of ohmic heating for milk pasteurization. This book discusses the overview of commercial thermal, non-thermal technologies and hybrid technologies for milk pasteurization. There are non-thermal technologies like pulse light, irradiation, ultra violet treatment etc., that can be used in combination with other technologies for the processing of milk and milk products. This hybrid technology can act as a potential area of research with multiple benefits, such as extended shelf life, reduced energy cost, reduced heat treatment, and better organoleptic and sensory properties. The book also describes the different aspects of food safety management used in a dairy processing unit. In the first section, the need of food safety management system is discussed, followed by risk analysis and different critical control points used in various dairy products that are discussed along with the benefits of FSMS.

Another new technology is microwave processing, which has several advantages as compared to the conventional methods used for thermal processing of milk. It can afford new challenges for shelf life extension with better quality retention, microbial stable product, and energy efficiency. The book also focuses on recent advances in microwave-assisted thermal processing of milk and the effects of microwaves on microbiological, physicochemical, and organoleptic properties of processed milk and milk products. Technological advances in value addition and standardization of the products have been reported, but a well-established process for mechanized production has been recommended in the book for obtaining a uniform quality nutritious product, produced under hygienic conditions.

**Vijendra Mishra, PhD**
Dean (Student Welfare) Associate Professor
(Microbiology) Department of Basic and Applied
Sciences National Institute of Food Tech.
Entrepreneurship & Management (NIFTEM), Plot
No. 97, Sector 56, HSIIDC Industrial Estate Kundli,
District- Sonipat, Haryana – 131028, India E-mail:
vijendramishra.niftem@gmail.com

March 2017

# WARNING/DISCLAIMER

# ABOUT THE SENIOR EDITOR-IN-CHIEF

**Megh R Goyal, PhD, PE**
*Retired Professor in Agricultural and Biomedical Engineering, University of Puerto Rico, Mayaguez Campus Senior Acquisitions Editor, Biomedical Engineering and Agricultural Science, Apple Academic Press, Inc.*

Megh R. Goyal, PhD, PE, is a Retired Professor in Agricultural and Biomedical Engineering from the General Engineering Department in the College of Engineering at University of Puerto Rico–Mayaguez Campus; and Senior Acquisitions Editor and Senior Technical Editor-in-Chief in Agriculture and Biomedical Engineering for Apple Academic Press Inc.

He has worked as a Soil Conservation Inspector and as a Research Assistant at Haryana Agricultural University and Ohio State University. He was the first agricultural engineer to receive the professional license in Agricultural Engineering in 1986 from the College of Engineers and Surveyors of Puerto Rico. On September 16, 2005, he was proclaimed as "Father of Irrigation Engineering in Puerto Rico for the twentieth century" by the ASABE, Puerto Rico Section, for his pioneering work on micro irrigation, evapotranspiration, agroclimatology, and soil and water engineering. During his professional career of 45 years, he has received many prestigious awards. A prolific author and editor, he has written more than 200 journal articles and textbooks and has edited over 50 books. He received his BSc degree in engineering from Punjab Agricultural University, Ludhiana, India; his MSc and PhD degrees from Ohio State University, Columbus; and his Master of Divinity degree from Puerto Rico Evangelical Seminary, Hato Rey, Puerto Rico, USA. Readers may contact him at: goyalmegh@gmail.com.

# ABOUT CO-EDITOR ANIT KUMAR

 Anit Kumar is PhD research scholar in the Department of Food Science and Technology at the National Institute of Food Technology Entrepreneurship and Management, Kundli, Haryana, India. He has reviewed research articles and review papers for the *Journal of Food Science and Engineering* and *Nutrition and Food Science*. He is a member of the Indian Dairy Association. He has working experience in the dairy and mango industry and also worked as a research scholar (Junior Research Fellow) at NIFTEM, Kundli, India, on the topic "Effect of microfluidization on the quality of fruit flavor and low fat yoghurt". He has also appeared on the NIFTEM YouTube where he demonstrated pedagogical and presentation skills through short films on the making process of patties, vanilla pastry, and puff khari. He has received the best organizer award as a convener for NIFTEM Sports League at NIFTEM, Haryana, India.

He completed his Bachelor of Technology in Food Process Engineering from SRM University, Chennai, India, and his Master of Technology in Food Engineering and Technology from SLIET, Punjab, India. During his undergraduate days, his aptitude in the study of machine designs to improve the production and processing of food led him to work on a five-month project in the dairy industry that was oriented toward the evaluation of process flow parameters. During his Master's thesis project, he worked upon improving the design of automatic 'pakoda' (an Indian snack) making machine. The project was intended to increase the overall hygiene of the food product while maintaining the taste.

He qualified GATE-2011 in XE-Food Technology and ICAR-NET 2015 in Food Technology. He has published several research articles and review papers in different international and national journals. He has also written several book chapters for different books. He has presented several oral and poster presentations at different international and national conferences. He received a best poster award as a co-author of "Enhancing quality of sugarcane juice: A non-thermal approach" at an international conference, "Food Value Chain: Innovations and Challenges—2016," held at NIFTEM, Haryana, India, during March, 2016. He also received third position among all the participants as a co-author of "Hybrid drying of onion slices using low pressure superheated steam and vacuum drying" at a national conference, "Technologies in Sustainable Food System—2016" held at SLIET, Punjab, India, during October, 2016.

# ABOUT CO-EDITOR ANIL K. GUPTA

 Anil K. Gupta, PhD, is the sole owner of AG Family Dentistry, LLC, in Alexandria, Louisiana. He was previously a Medical Technologist at Huey P. Long Hospital, Pineville, Louisiana. Dr. Gupta, after completing his PhD (1994) from Haryana Agricultural University, worked in TERI (TATA Energy Research Institute, India) as a research associate, where he published several papers on soil and water. After coming to US in 1995, he retrained himself as a Medical Technologist. Dr. Gupta started working at the Huey P. Long Hospital, Pineville, Louisiana (USA), as a Medical Technologist in 1998 and worked for about ten years. While his stay there, Dr. Gupta decided to go to direct patient care and went to dental school for five years. After graduating in 2014, he worked with a dentist for a year and then opened his own practice in June 2015. Today, Dr. Gupta is a proud owner of a dental practice, which is reaching new heights.

# OTHER BOOKS ON AGRICULTURAL & BIOLOGICAL ENGINEERING BY APPLE ACADEMIC PRESS, INC.

**Management of Drip/Trickle or Micro Irrigation**
Megh R. Goyal, PhD, PE, Senior Editor-in-Chief

**Evapotranspiration: Principles and Applications for Water Management**
Megh R. Goyal, PhD, PE, and Eric W. Harmsen, Editors

**Book Series: Research Advances in Sustainable Micro Irrigation**
Senior Editor-in-Chief: Megh R. Goyal, PhD, PE

Volume 1:  Sustainable Micro Irrigation: Principles and Practices
Volume 2:  Sustainable Practices in Surface and Subsurface Micro Irrigation
Volume 3:  Sustainable Micro Irrigation Management for Trees and Vines
Volume 4:  Management, Performance, and Applications of Micro Irrigation Systems
Volume 5:  Applications of Furrow and Micro Irrigation in Arid and Semi-Arid Regions
Volume 6:  Best Management Practices for Drip Irrigated Crops
Volume 7:  Closed Circuit Micro Irrigation Design: Theory and Applications
Volume 8:  Wastewater Management for Irrigation: Principles and Practices
Volume 9:  Water and Fertigation Management in Micro Irrigation
Volume 10: Innovation in Micro Irrigation Technology

**Book Series: Innovations and Challenges in Micro Irrigation**
Senior Editor-in-Chief: Megh R. Goyal, PhD, PE

- Micro Irrigation Engineering for Horticultural Crops: Policy Options, Scheduling and Design
- Micro Irrigation Management: Technological Advances and Their Applications

- Micro Irrigation Scheduling and Practices
- Performance Evaluation of Micro Irrigation Management: Principles and Practices
- Potential of Solar Energy and Emerging Technologies in Sustainable Micro Irrigation
- Principles and Management of Clogging in Micro Irrigation
- Sustainable Micro Irrigation Design Systems for Agricultural Crops: Methods and Practices
- Engineering Interventions in Sustainable Trickle Irrigation: Water Requirements, Uniformity, Fertigation, and Crop Performance

**Book Series: Innovations in Agricultural & Biological Engineering**
Senior Editor-in-Chief: Megh R. Goyal, PhD, PE

- Dairy Engineering: Advanced Technologies and Their Applications
- Developing Technologies in Food Science: Status, Applications, and Challenges
- Engineering Interventions in Agricultural Processing
- Engineering Practices for Agricultural Production and Water Conservation: An Inter-disciplinary Approach
- Emerging Technologies in Agricultural Engineering
- Flood Assessment: Modeling and Parameterization
- Food Engineering: Emerging Issues, Modeling, and Applications
- Food Process Engineering: Emerging Trends in Research and Their Applications
- Food Technology: Applied Research and Production Techniques
- Modeling Methods and Practices in Soil and Water Engineering
- Processing Technologies for Milk and Dairy Products: Methods Application and Energy Usage
- Soil and Water Engineering: Principles and Applications of Modeling
- Soil Salinity Management in Agriculture: Technological Advances and Applications
- Technological Interventions in the Processing of Fruits and Vegetables
- Technological Interventions in Management of Irrigated Agriculture
- Engineering Interventions in Foods and Plants

- Technological Interventions in Dairy Science: Innovative Approaches in Processing, Preservation, and Analysis of Milk Products
- Novel Dairy Processing Technologies: Techniques, Management, and Energy Conservation
- Sustainable Biological Systems for Agriculture: Emerging Issues in Nanotechnology, Biofertilizers, Wastewater, and Farm Machines
- State-of-the-Art Technologies in Food Science: Human Health, Emerging Issues and Specialty Topics
- Scientific and Technical Terms in Bioengineering and Biological Engineering

# EDITORIAL

---

Apple Academic Press Inc., (AAP) is publishing various book volumes on the focus areas under the book series titled *Innovations in Agricultural and Biological Engineering*. Over a span of 8 to 10 years, Apple Academic Press Inc., will publish subsequent volumes in the specialty areas defined by *American Society of Agricultural and Biological Engineers* (<asabe. org>).

The mission of this series is to provide knowledge and techniques for agricultural and biological engineers (ABEs). The series aims to offer high-quality reference and academic content in agricultural and biological engineering (ABE) that is accessible to academicians, researchers, scientists, university faculty, and university-level students and professionals around the world. The following material has been edited/ modified and reproduced below [From: *Goyal, Megh R., 2006. Agricultural and biomedical engineering: Scope and opportunities. Paper Edu_47 Presentation at the Fourth LACCEI International Latin American and Caribbean Conference for Engineering and Technology (LACCEI' 2006): Breaking Frontiers and Barriers in Engineering: Education and Research by LACCEI University of Puerto Rico – Mayaguez Campus, Mayaguez, Puerto Rico, June 21 – 23*]:

## WHAT IS AGRICULTURAL AND BIOLOGICAL ENGINEERING (ABE)?

*"Agricultural Engineering (AE) involves application of engineering to production, processing, preservation and handling of food, fiber, and shelter. It also includes transfer of technology for the development and welfare of rural communities"*, according to <isae.in>. *"ABE is the discipline of engineering that applies engineering principles and the fundamental concepts of biology to agricultural and biological systems and tools, for the safe, efficient and environmentally sensitive produc-tion, processing, and management of agricultural, biological, food, and*

*natural resources systems"*, according to <asabe.org>. *"AE is the branch of engineering involved with the design of farm machinery, with soil management, land development, and mechanization and automation of livestock farming, and with the efficient planting, harvesting, storage, and processing of farm commodities"*, definition by: <http://dictionary.reference.com/browse/agricultural+engineering>.

*"AE incorporates many science disciplines and technology practices to the efficient production and processing of food, feed, fiber and fuels. It involves disciplines like mechanical engineering (agricultural machinery and automated machine systems), soil science (crop nutrient and fertilization, etc.), environmental sciences (drainage and irrigation), plant biology (seeding and plant growth management), animal science (farm animals and housing) etc.,"* by <http://www.ABE.ncsu.edu/academic/agricultural-engineering.php>.

According to https://en.wikipedia.org/wiki/Biological_engineering: *"BE (Biological engineering) is a science-based discipline that applies concepts and methods of biology to solve real-world problems related to the life sciences or the application thereof. In this context, while traditional engineering applies physical and mathematical sciences to analyze, design and manufacture inanimate tools, structures and processes, biological engineering uses biology to study and advance applications of living systems."*

## SPECIALTY AREAS OF ABE

Agricultural and Biological Engineers (ABEs) ensure that the world has the necessities of life including safe and plentiful food, clean air, and water, renewable fuel and energy, safe working conditions, and a healthy environment by employing knowledge and expertise of sciences, both pure and applied, and engineering principles. Biological engineering applies engineering practices to problems and opportunities presented by living things and the natural environment in agriculture. BA engineers understand the interrelationships between technology and living systems, have available a wide variety of employment options. The <asabe.org> indicates that *"ABE embraces a variety of following specialty areas"*. As new technology and information emerge, specialty areas are created, and many overlap with one or more other areas.

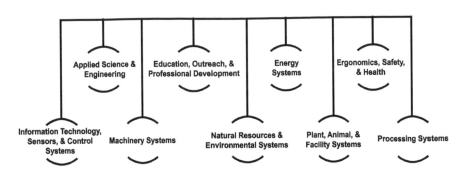

1. **Aquacultural Engineering**: ABEs help design farm systems for raising fish and shellfish, as well as ornamental and bait fish. They specialize in water quality, biotechnology, machinery, natural resources, feeding and ventilation systems, and sanitation. They seek ways to reduce pollution from aquacultural discharges, to reduce excess water use, and to improve farm systems. They also work with aquatic animal harvesting, sorting, and processing.

2. **Biological Engineering** applies engineering practices to problems and opportunities presented by living things and the natural environment. It also includes applications of nanotechnology in agricultural and biological systems.

3. **Energy:** ABEs identify and develop viable energy sources—biomass, methane, and vegetable oil, to name a few—and to make these and other systems cleaner and more efficient. These specialists also develop energy conservation strategies to reduce costs and protect the environment, and they design traditional and alternative energy systems to meet the needs of agricultural operations.

4. **Farm Machinery and Power Engineering:** ABEs in this specialty focus on designing advanced equipment, making it more efficient and less demanding of our natural resources. They develop equipment for food processing, highly precise crop spraying, agricultural commodity and waste transport, and turf and landscape maintenance, as well as equipment for such specialized tasks as removing seaweed from beaches. This is in addition to the tractors, tillage equipment, irrigation equipment, and harvest equipment that have done so much to reduce the drudgery of farming.

5. **Food and Process Engineering:** Food and process engineers combine design expertise with manufacturing methods to develop economical and responsible processing solutions for industry. Also food and process engineers look for ways to reduce waste by devising alternatives for treatment, disposal and utilization.

6. **Forest Engineering:** ABEs apply engineering to solve natural resource and environment problems in forest production systems and related manufacturing industries. Engineering skills and expertise are needed to address problems related to equipment design and manufacturing, forest access systems design and construction; machine–soil interaction and erosion control; forest operations analysis and improvement; decision modeling; and wood product design and manufacturing.

7. **Information and Electrical Technologies Engineering** is one of the most versatile areas of the ABE specialty areas, because it is applied to virtually all the others, from machinery design to soil testing to food quality and safety control. Geographic information systems, global positioning systems, machine instrumentation and controls, electromagnetics, bioinformatics, biorobotics, machine vision, sensors, spectroscopy: These are some of the exciting information and electrical technologies being used today and being developed for the future.

8. **Natural Resources:** ABEs with environmental expertise work to better understand the complex mechanics of these resources, so that they can be used efficiently and without degradation. ABEs determine crop water requirements and design irrigation systems. They are experts in agricultural hydrology principles, such as controlling drainage, and they implement ways to control soil erosion and study the environmental effects of sediment on stream quality. Natural resources engineers design, build, operate and maintain water control structures for reservoirs, floodways, and channels. They also work on water treatment systems, wetlands protection, vertical farming and other water issues.

9. **Nursery and Greenhouse Engineering**: In many ways, nursery and greenhouse operations are microcosms of large-scale production agriculture, with many similar needs—irrigation, mechanization, disease and pest control, and nutrient application. However, other engineering needs also present themselves in nursery and greenhouse operations: equipment for transplantation; control systems for temperature, humidity, and ventilation; and plant biology issues, such as hydroponics, tissue culture, and seedling propagation methods. And sometimes the challenges are extraterrestrial: ABEs at NASA are designing greenhouse systems to support a manned expedition to Mars!

10. **Safety and Health:** ABEs analyze health and injury data, the use and possible misuse of machines, and equipment compliance with standards and regulation. They constantly look for ways in which the safety of equipment, materials and agricultural practices can be improved and for ways in which safety and health issues can be communicated to the public.

11. **Structures and Environment:** ABEs with expertise in structures and environment design animal housing, storage structures, and greenhouses, with ventilation systems, temperature and humidity controls, and structural strength appropriate for their climate and purpose. They also devise better practices and systems for storing, recovering, reusing, and transporting waste products.

## CAREER IN AGRICULTURAL AND BIOLOGICAL ENGINEERING

One will find that university ABE programs have many names, such as biological systems engineering, bioresource engineering, environmental engineering, forest engineering, or food and process engineering. Whatever the title, the typical curriculum begins with courses in writing, social sciences, and economics, along with mathematics (calculus and statistics), chemistry, physics, and biology. Student gains a fundamental knowledge of the life sciences and how biological systems interact with their environment. One also takes engineering courses, such as thermodynamics, mechanics, instrumentation and controls, electronics and electrical circuits, and engineering design. Then student adds courses related to particular interests, perhaps including mechanization, soil and water resource management, food and process engineering, industrial microbiology, biological engineering or pest management. As seniors, engineering students work in a team to design, build, and test new processes or products.

For more information on this series, readers may contact:

Ashish Kumar, Publisher and President
Sandy Sickels, Vice President
Apple Academic Press, Inc.
Fax: 866-222-9549
E-mail: ashish@appleacademicpress.com
http://www.appleacademicpress.com/
publishwithus.php

Megh R. Goyal, PhD, PE
Book Series Senior
Editor-in-Chief
*Innovations in Agricultural
and Biological Engineering*
E-mail: goyalmegh@gmail.
com

# PART I

# Emerging Processing Technologies in the Dairy Industry

# CHAPTER 1

# TECHNOLOGY OF OHMIC HEATING FOR THE PASTEURIZATION OF MILK

ASAAD REHMAN SAEED AL-HILPHY[1*], HAIDER I. ALI[1], and GHASSAN F. MOHSIN[2]

[1]*Department of Food Science, College of Agriculture, University of Basrah, Basrah City, Iraq*

[2]*Vocational School, Maysan City, Iraq*

*Corresponding author. E-mail: aalhilphy@yahoo.co.uk*

## CONTENTS

This chapter includes information from "*Al-Hilphy, A. R. S.; Haider, I. A.; Mohsin, G. F. Designing and manufacturing milk pasteurization device by ohmic heating and studying its efficiency. Journal of Basrah Researches (Sciences): Academic Scientific J. (Basrah University Faculty)*, 2012, *4(38)*, *1-18*", which is an open access article.

## ABSTRACT

The best voltage for milk pasteurization by using ohmic heating was 80 V compared with 110, 220 V, both of which were inadequate for pasteurization. Heating at 80 V was safer and inexpensive as it required less power (80 V) in contrast with higher voltages (110–220 V) and less dangerous. Voltage of 80 V does not change milk color to brownish as it was noticed at 110– 220 V and undesirable smells were not present. Heating at 80 V did not cause fouling or precipitation of deposits on electrodes compared to 110–220 V. The best electrical conductivity was at 80 V as there were no deposits on the electrodes. The practical temperature of milk was stable because of the thermal valve, which controls the temperature at 72°C. The electric conductivity and current were increased with increasing temperature in ohmic heating at 220, 110, and 80 V while these were decreased with the increase of temperature at 220 V. The thermal conductivity and thermal diffusivity were increased with the increase of temperature at all voltages in the ohmic heating and high time short temperature (HTST).

The viscosity of milk and its density were decreased with the increase of temperature at all voltages in the ohmic heating and HTST. The period of keeping milk in the device was decreased with the increase of voltage in the ohmic heating, which was less than HTST. At 80 V of treatment, the highest coefficient of performance was 0.80, compared with 220, 110 V, with performance coefficient of 0.49 and 0.76, respectively. The percentages of protein, lipid, lactose, ash, and humidity in raw milk were 3.6, 3.7, 5.02, 0.68, and 87.0%, respectively. These values at 80 V were 3.5, 3.6, 6.2, 0.69, and 85.9%, respectively; and at 110 V were 3.54, 3.6, 6.9, 0.69, and 85.2%, respectively; and at 220 V were 3.5, 3.6, 7.1, 0.73, and 85.0%, respectively, and in HTST were 3.57, 3.7, 6.0, 0.71, and 87.0%, respectively.

The percentage of acidity in raw milk was 0.15% and was 0.14, 0.14, 0.13, and 0.15%, respectively, for ohmic heating at 220, 110, 80 V, and HTST. The pH prior the pasteurization was 6.6 compared to 6.7, 6.8, 6.8, and 6.8 after the pasteurization by ohmic heating at 220, 110, 80 V, and HTST treatment, respectively. The phosphatase enzyme test gave positive results in raw milk and negative one in pasteurized milk treated either by ohmic heating on all voltages and HTST. The clot on boiling test and turbidity test gave negative results for raw milk. The microbiological results of pasteurized milk by ohmic heating showed no colonies in total count bacteria, coli form, Staph-110, yeasts, and molds. Milk treated by HTST pasteurization had bacteria while bacteria were absent. The shelf-life study of pasteurized milk by ohmic heating at 4°C for 15 days showed no change in pH and acidity. The results also showed no colonies in bacterial total count bacteria, coli form, staph-110, yeasts, and molds. However, pasteurized milk by both methods (ohmic and HTST) could only be stored for 72 h at room temperature. The HTST treated milk could be stored for 8 days at 4°C for 8 days.

## 1.1   INTRODUCTION

Since dairy products contain most of the nutrients needed to build the human body, milk contains a number of nutrients such as carbohydrates, fats, proteins, minerals, and vitamins besides water. These nutrients distinguish milk to have a variety of nutritional values resulting in highest biological value in comparison with other foods.[6]

It is nearly impossible to have milk with no microorganism even though there is availability of ideal healthy conditions and sanitation of animals or milkman. Contamination includes mixed microorganisms. Part of it is harmful to human health as these produce lactic acid. Others are harmful because these contribute in transporting disease or produce poisons. Therefore, milk can be a source of spreading diseases. It has been found that the thermal treatment to the milk leads to the preservation of physiological properties of milk. Also, the milk without adequate treatments in the dairy industry is not of good quality.[8] The most common thermal treatments used by dairy industry are pasteurization and sterilization. The common properties of pasteurization of milk focus on two aspects:

- Health aspect, which eliminates pathogenic organisms in milk, eliminates 95–99% of bacteria in milk as well as 100% of yeasts and molds due to milk exposition to different temperatures at different times.
- The second aspect increases the ability of preserving milk due to the elimination of pathogenic organisms.[2]

United States Food and Drug Administration (USFDA) has introduced standards for pasteurizing milk and milk products that includes treating and exposing the milk to: 62.8°C for 30 min, 71.6°C for 15 min, 88.4°C for 0.1 s, 95.6°C for 0.05 s, and 100°C for 0.01 s.[27]

Al-Dehan[2] indicated three methods of milk pasteurization:

- **Batch pasteurization**: In this type of pasteurization, milk is exposed to 62.8°C for a period not less than 30 min. The treatment is done in batches and is also called dock (tank) method.
- **High temperature short time (HTST)**: Milk is exposed to 72°C for 15 s.
- **Vacuums pasteurization**: Milk is exposed to different thermal treatments under vacuum. The objective is to get rid of bad smell of some dairy products. Milk is exposed to 90.5–96.1°C under vacuum pressure.

There are other methods of pasteurization such as:

- **Continuous-flow system:** Milk is allowed to flow through pipes and at the same time it is exposed to different thermal treatments for half an hour and the length of pipes ensures adequate thermal treatment.
- **Infra-red treatment:** Milk is allowed to flow through horizontal rustproof iron pars that are exposed to infrared rays until temperature rises to 85°C for a desired period.[49]
- **Microwave:** Milk is exposed to the microwave energy in batches to have a temperature of 72°C for 15 s.[4]
- Use of solar energy.[1]

The ohmic heating to treat foods is being widely used throughout the world, due to low requirements of electricity and the need of small space in comparison with traditional methods.[75] Also ohmic heating has shown benefits over other methods including coordination and regulation in heating and energy efficiency by giving it a shortcut in the design and lack of phenomenon of slowdown.[74] The use of electrical technology in preparing food gives clean and efficient energy suitable to environment compared with classical methods.[81] Recently research studies have focused on the design and development of efficiency of ohmic heating for food and sterilization at Mars exploration vehicle.[41,42]

This chapter focuses on the technology of ohmic heating for milk pasteurization.

## 1.2 HISTORY OF PASTEURIZATION

The term "Pasteurization" was named due to famous French chemical scientist Lewis Pasteur (1822–1895), who had established manufacturing microbial science rules. During the years 1864–1865 and 1871–1872, Pasteur had treated alcoholic drinks by 50–60°C to eliminate unwanted fermenting microbes.[17] He concluded that milk fermentation was similar to alcoholic fermentation, as they shared the same fermenting factor, in growth of unwanted microbes. In 1880, he started the use of pasteurizing devices to heat the milk at the temperature range of 60–70°C without determining the retention time. However, he rapidly paid attention to the fact that milk pasteurized with no retention time could cause diseases to consumers transmitted by milk. At the end of that century, scientists started the use of pasteurizing devices using pipes to seize milk at a pasteurization temperature. They found that suitable temperature was 60°C for 20 min for milk free from *Mycobacterium tuberculosis* bacteria.[27]

In 1935, it was found that pasteurized milk had alkaline phosphatase enzyme as an indicator so that pasteurization was not efficient as it inhibits this enzyme.[21]

HTST method is an effective method,[27] as it heats every molecule of milk to the desirable temperature (71.1°C for 15 s). This method eliminates 95–99% of bacteria in milk and extends its storage life remarkably.[60] Pasteurization has been practically used for decades as a method to preserve liquid milk for longer periods.[48,66,86] Pasteurizing milk by HTST method can extend ideal preserving period for six months without fridge,[66] though some changes may occur during storage period as gelatin formation.

## 1.3 MILK PASTEURIZATION

Milk pasteurization is a thermal treatment to each molecule of milk at a temperature less than 100°C. It can be done either by slow method as temperature rises between 62 and 65°C for 30 min or by HTST method with a temperature of 71.8°C for 15–40 min or by flash heating (85–90°C for 1–4 s) or exposing milk to a temperature between 94°C and 100°C for 0.01–0.1 s.[25,50] There are other methods like: microwave with different batches[3] or solar heating.[1]

## 1.4 DEFINITION OF ELECTRICITY

Electricity consists of concepts less popular than electromagnetic field and electromagnetic induction. Electrical power is more flexible type of power source at low cost in comparison with other sources of energy.

The word electricity consists of two Persian words; "*kah*" means straw and "rby" means magnetic and electricity means "*Kahrman*" (amber) in Persian. The Latin word is "electricus," which means like amber. The latest word also taken from Greek word ἤλεκτρον (electron) also means amber.[12] This association gives two words "electric" and "electricity," which was first shown by Thomas Brown book "Common Mistakes" in Latin "*Pseudodixa Epidemica*" edited in 1646.[19]

Although early 19th century had shown fast development in electricity science, yet the late period of this century showed most of the developments in electrical engineering. Therefore, electricity changed from the scientific curiosity to main indispensable tool in modern life and the driving force to second industrial revolution. This rapid development is owed to the work of renowned scientists, such as Nikola Tesla, Thomas Edison, Otto Bly, George Westinghouse, Ernst Von Siemens, Alexander Graham Bell, Lord William Watson, and Barron Kelvin I. Electricity is very flexible type of energy, as it suits any type of use.[16,68,94]

In the 18th century, a German scientist, George Simon Ohmic, drafted a law called Ohmic's law. During his studies, Ohmic proved that the electric current is directly proportional to the voltage on the circuit and the relationship of the current and the voltage in the circuit is linear as well as power is inversely proportional to the value of the total resistance of the circuit as shown in Figure 1.1.

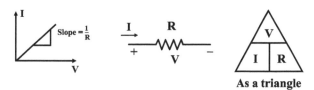

**FIGURE 1.1** The relationship between the current and voltage.

Ohmic law states that the current ($I$) flowing in the resistance is directly proportional to the applied voltage ($V$) and inversely to the resistance ($R$) as described by the following equation:[71]

$$I = V/R \qquad\qquad (1.1)$$

The different forms of Ohmic formula[9,38] are:

$$I = V/R \qquad\qquad (1.2)$$

$$R = V/I \text{ and } V = I \times R \qquad\qquad (1.3)$$

It can be concluded that name and mechanical technique of ohmic heating was taken from Ohmic's law depending on the voltage difference and resistance. The success of using electricity in food processing was developed in 19th century, when milk pasteurization was introduced.[30] This method of pasteurization is called electropure process.

In 1938, five states of USA used method of pasteurization and served about 5000 consumers.[52] In the 20th century, milk was pasteurized by passing electricity through parallel plates. During the past two centuries, new and developed materials and different designs of ohmic heating technique have been introduced. Great Britain Electricity Council (GBEC) granted the patent on continuous flow in ohmic heating and authorized its use to APV Baker Company.[84] This technique was left as a result of high cost.[23] Finally ohmic heating got more attention because of best quality of the product compared to other classical techniques.[18,44,58,75] Ohmic heating technique has also been called *Joule or resistive heating*. In this method, electric current passes through the food material causing a homogeneous distribution of heat inside the food.[35,46,81] The fast work, homogenized heat and less loss of food vitamins are characteristics of ohmic method.[91,92,97]

Increase in homogeneous heat of food gives well-accepted products.[27] The advantages of this method are: simple to use, high-energy efficiency, and low cost compared to microwave and radio vibration methods.[31,44,51] Processing of foods by electric technique is environment friendly.[43,81] The use of ohmic heating gives a product of acceptable quality.[47]

There are several precautions to be taken into consideration while designing the pasteurization device. One of these precautions is the ability to heat to a recommended pasteurization temperature, as the temperature should be chosen before designing the heat exchanger. Further, pasteurization temperature depends on the seizing period and pasteurization of nonnaturalized milk. Therefore, the heat exchanger should be able to heat the milk to a temperature around 73–74°C, which is clearly greater than the desirable pasteurization temperature.[27] Other precaution is the ability to keep the product at a pasteurized temperature to get rid of pathogenic microbes, inhibition of alkaline phosphatase enzyme, cooling ability, controlling pasteurization temperature, working accuracy, and the ability to have minimum changes in chemical and organoleptic properties of milk, pasteurization device free from microbial and chemical contamination.[27] The design of ohmic heating technique must be taken into consideration: type of product to be pasteurized, its characteristics and properties, electrical conductivity, and heating coefficient of this substrate.[47]

Ohmic heating device by Ayadi et al.[11] consisted of five ohmic cells in the form of rectangular channels (cell length of 240 mm, thickness of 15 mm, and width of 75 mm): three cells for heating and the other two for electrical isolation. Japan started production of ohmic heating system in 1995,[53] while Yanagiya developed ohmic heating system type MINI-J with titanium poles in 1999.[53] Ghnimi et al.[31] designed an ohmic heating system (Fig. 1.2) with feed pump type PCM Moineau, vibratory tank and two heat exchangers (the first from pipe type double cooling and the second was an ohmic cell for heating). Kong et al.[45] designed ohmic heating device consisting: ohmic heating unit that included voltage exchanger device and pores from rust resistance steel and teflon pipe; and data collection system consisting of a digital conductivity meter and thermocouple.

Samaranayake and Sastry[72] indicated that the use of poles of stainless steel is one of the very active electrochemical substances during ohmic heating to resist all pH values. Stirling[87] and Berthou et al.[15] used titanium poles covered with platinum or Raytheon to inhibit electrical that occurs during alternating current with a low frequency (50 or 60 Hz). Reznik[70] used stainless steel poles for a frequency >100 Hz or poles from graphite. Reznik[69] added that electrolyte between electrical pole and the product could prevent the product from contamination.

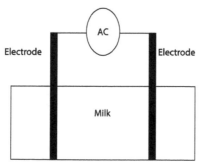

**FIGURE 1.2** Ohmic heating system.

## 1.5   DESIGN OF OHMIC MILK PASTEURIZER

Ohmic milk pasteurizer (Figs. 1.3 and 1.4) consists of raw milk tank (double jacket of 25 L capacity made of stainless steel), feed pump of plastic of 70 W power, heat exchanger of 120 cm in length and 1.25 cm diameter, heating tube made of heat resisting Teflon of 36 cm length and 5 cm diameter; and consists of electrodes made of stainless steel that can heat the milk at three different voltages 80, 110, and 220 V.

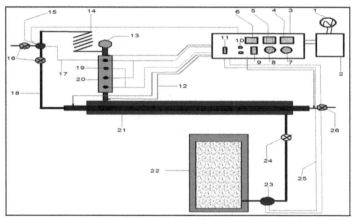

**FIGURE 1.3**   Layout of milk ohmic pasteurizer.[5]

**FIGURE 1.4**   Image of milk ohmic pasteurizer.[5]

These electrodes were organized in the tube in a parallel arrangement and the distance between these was of 5.5 cm. The device also includes a holding tube of 4 m in length and 1 cm in diameter; to hold milk for 15 s at 72°C. It also has an electric valve, nonreverse delivery valve and manual valve. The design equations are presented below:

Overall heat transfer (U) is calculated as following:[25]

$$\frac{1}{U} = \frac{1}{h_a} + \frac{X}{K} + \frac{1}{h_b}$$ (1.4)

where $h_a$, $h_b$, $X$, and $K$ are heat transfer by convection for hot and cold milk (W/m.°C), pipe thickness and heat transfer by conduction for pipe (W/m².°C), respectively.

$$R_e = \frac{DV\rho}{\mu}$$ (1.5)

$$N_u = \frac{h_a D}{K}$$ (1.6)

$$P_r = \frac{C_p \mu}{K}$$ (1.7)

$$N_u = 1.62 \, (R_e P_r \frac{D}{L})^{0.33}$$ (1.8)

$$h_u = 1.62 \frac{K}{D} \, (R_e P_r \frac{D}{L})^{0.33}$$ (1.9)

where $R_e$, $N_u$, $P_r$, $\mu$, $D$, $C_p$, and $L$ are Reynolds number, Nusslet number, Prandtl number, milk viscosity (Pa.s), pipe diameter, milk specific heat, and pipe length (m), respectively. Physical properties of milk were taken at mean bulk temperature.

Mean logarithmic difference of temperatures $\Delta T_m$ is calculated from the following equation:

$$\Delta T_m = \frac{(T_{hi-}T_{co}) - (T_{ho-}T_{ci})}{\ln(\frac{T_{hi-}T_{co}}{T_{ho-}T_{ci}})}$$ (1.10)

$$q = mc_p \left( T_{co} - T_{ci} \right) \tag{1.11}$$

$$q = U A \, \Delta T_m \tag{1.12}$$

$$q / U \Delta T_m \tag{1.13}$$

$$A = \pi D L \tag{1.14}$$

$$L = A / \pi D \tag{1.15}$$

where $q$, $T_{hi}$, $T_{ho}$, $T_{co}$, and $T_{ci}$ are heat energy, hot milk temperature entering into heat exchanger (°C), hot milk temperature exiting from heat exchanger (°C), heated milk temperature exiting from heat exchanger and going to heating unit (°C), and cold milk temperature entering into heat exchanger (°C).

Ratio of regeneration heat in the heat exchanger was 65%. Ratio of regeneration heat is calculated from the following equation:

$$\varepsilon_h = \frac{T_{co} - T_{ci}}{T_{hi} - T_{ci}} \times 100 \tag{1.16}$$

Milk temperature that exit from ohmic heating unit is calculated from the following equation:[14]

$$\frac{aT + b}{aT_o + b} = e^{\left| \frac{a \pi d_c L}{m' c_p} \right|} \tag{1.17}$$

$$a = \frac{\left| \Delta V \right|^2 d_c \sigma_{o\,m^n}}{4} - U \tag{1.18}$$

$$b = \frac{d_c \left| \Delta V \right|^2 \sigma_o}{4} + U \tag{1.19}$$

where $(\Delta V)$, $\sigma_o$, $d_c$, and $L$ are voltage progress on the long of heating tube (V/m), electrical conductivity for milk at temperature of 0°C, heating tube diameter (m$^2$), and heating tube length (m), respectively. $m^n$ is the constants related to electrical conductivity[56] according to the following equation:

$$\sigma_L = \sigma_o \left( 1 + m^n \, T \right) \tag{1.20}$$

where $\sigma_L$, $m'$, and $U$ are milk electrical conductivity (S/m), mass flow rate of milk (kg/s) and over all heat transfer of milk (W /m$^2$.°C) depends on outside area of heating tube (m$^2$), respectively.

$$q = h_o A (T_{ins} - T_a) \tag{1.21}$$

$$h_o = \frac{q}{A(T_{ins.} - T_a)} \tag{1.22}$$

where $q$, $h_o$, $T_{ins}$, $T_a$, and $A$ are lost heat energy (W), heat transfer by convection between insulator and ambient (W /m$^2$.°C), insulator temperature (°C), ambient temperature (°C), and insulator area (m$^2$), respectively.

Heat transfer coefficient by convection ($hi$) is calculated using Nusslet number as follows:

$$N_{Nu} = \frac{h_i D}{K} \tag{1.23}$$

where $N_{Nu}$, $K$, and $D$ are Nusselt number, heat transfer coefficient by conduction (W /m.°C), and diameter (m), respectively.

Mean temperature of milk for determination of physical and heat properties of milk is calculated as follows:

$$T_f = \frac{T_{in} + T_w}{2} \tag{1.24}$$

$$N_{Nu} = 3.66 + \frac{0.085 \left( N_{Re} \times N_{Pr} \times \dfrac{D}{L} \right)}{1 + 0.045 \left( N_{Re} \times N_{Pr} \times \dfrac{D}{L} \right)} (\frac{\mu_b}{\mu_w})^{0.14} \tag{1.25}$$

where $\mu_b$ and $\mu_w$ are milk viscosity (pa.s), and milk viscosity at mean temperature (pa.s.), respectively.

$$q = h_i A (T_{in} - T_w) \tag{1.26}$$

where $T_w$ and $T_{in}$ are milk temperature entering pipe (°C) and milk temperature exit of pipe (°C), respectively.

$$q = V^2 . A\sigma_L / L \tag{1.27}$$

$$U = \cfrac{1}{\cfrac{r_3}{r_{1h_i}} + \cfrac{r_3 ln\left(\frac{r_2}{r_1}\right)}{K_A} + \cfrac{r_3 ln\left(\frac{r_3}{r_2}\right)}{K_B} + \cfrac{1}{h_o}} \qquad (1.28)$$

where $r_1$, $r_2$, $r_3$, $K_A$, and $K_B$ are internal radius of cylinder (m), external radius of cylinder (m), total radius with insulator (m), and heat transfer coefficient by (W/m.°C), respectively.

On the other hand, the length of holding tube is calculated by the following equation:[89]

$$L_H = \frac{4V}{\pi D^2} \qquad (1.29)$$

$$V = \frac{QHT}{3600\,\eta} \qquad (1.30)$$

where $Q$, $HT$, $L_H$, $D$, $V$, and $\eta$ are mean flow rate of milk (m³/h), holding time (s), holding tube length (m), internal diameter of pipe (m), milk volume during $Q$ and $HT$, and efficiency factor (= 0.85), respectively.

## 1.6   EFFECTS OF OHMIC HEATING ON PHYSICAL PROPERTIES OF MILK

### 1.6.1   ELECTRICAL CONDUCTIVITY

Electrical conductivity is one of the important factors which affect the success of ohmic heating technique. Conductivity increases with mineral content, acids, and moisture; and reduces with fat and alcohol present in the food. Electrical conductivity can be measured based on equations by Wang and Sastry[96] and Icier et al.:[36]

$$\sigma = \frac{IL}{VA} \qquad (1.31)$$

where $I$ = current (A), $L$ = distance between poles (m), $V$ = voltage (V), $A$ = area (m²), and $\sigma$ = the electrical conductivity (S/m).

The ability of material to conduct electricity is called specific resistance, which is known as solution pole resistance (R, Ohms) for 1 cm in length and 1 cm thickness of sample. The specific conductance is a reciprocal of specific resistance, and it is very low (0.005) for natural milk but it increases for skim milk from infected mammary gland (mastitis) because of high level of sodium and chlorine.[8] However, Alnimer[6] indicated that fresh milk conductivity ranges from $45 \times 10^4$ to $48 \times 10^4$ S/m and increases to reach $10–13 \times 10^5$ S/m when there is mastitis because of high chloride, which shows an importance of measuring conductivity.

Milk has a homogenous electrical conductivity due to enough free water saturated with solved ionic minerals, which makes the milk suitable for ohmic treatment.[55] Novy and Zinty[54] found that milk conductivity was increased with increase in heat, and was 0.5, 0.65, 0.94, and 1.1 S/m at a temperature of 20, 40, 60, and 80°C, respectively.

Figure 1.5[5] shows that electrical conductivity was increased with increasing milk temperature in the high traditional pasteurization and ohmic heating at 110 and 80 V; while at 220 V, the electrical conductivity was decreased with increasing milk temperature. Also, the Figure 1.5 shows that electrical conductivity at 220 V reached to 0.4 S/m at 22°C then was reduced with increasing temperature till it was 0.25 S/m. Many researchers have found that the electrical conductivity of food products always increases with increasing temperature, food moisture content, and ionic salts by ohmic heating at 220 V, implying that heating is desirable

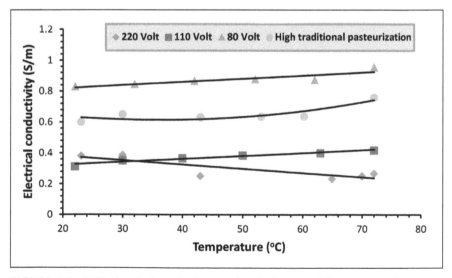

**FIGURE 1.5**   Milk electrical conductivity under ohmic pasteurizing treatment.[5]

for treatment at high moisture content (87%) and ionic salts percentage of 0.7%. The high reduction in the electrical conductivity was because of presence of deposits on the electrodes.

Electrical conductivity of foods changes with temperature depending on the food type. In some foods, the electrical conductivity increases with increasing temperature with ohmic heating at 220 V, because the proteins do not denature on the electrodes, but it causes denaturation of whey proteins that is affected by high temperature. The relationship between electrical conductivity at 220 V and milk temperature is a linear equation with an intercept of $-0.0028$ as shown below:

$$Ec_{220V} = -0.0028\ T + 0.4351 \tag{1.32}$$

$$Ec_{110V} = +0.001\ T + 0.285 \tag{1.33}$$

$$Ec_{80V} = +0.002\ T + 0.778 \tag{1.34}$$

However, the relationship between electrical conductivity and temperature at 110 V is linear equation with a positive slope and with a determination coefficient of 0.94. It may be due to reduction of formation of deposits on the electrodes, at low voltage.

The electrical conductivity reached 0.8 S/m at 80 V and no fouling occur on the electrodes. Equation 1.34 indicates the relationship between the electrical conductivity and temperature.

In the high traditional pasteurization, the electrical conductivity reached to 0.6 S/m at temperature of 22°C then was increased with increasing milk temperature till temperature reached 72°C. This is because a part of moisture content was evaporated from milk that led to increase in concentration of salts in milk, which caused the increase in electrical conductivity. The following equation shows the relationship between electrical conductivity and milk temperature:

$$Ec_{Cont} = 1 \times 10^{-4}\ T^2 - 0.0071\ T + 0.7405 \tag{1.35}$$

## 1.6.2 ELECTRICAL CURRENT

Figure 1.6 shows that electrical current was increased with increasing milk temperature at 110 and 80 V, while it was decreased with temperature at 220 V because of formation of fouling on the electrodes.[5]

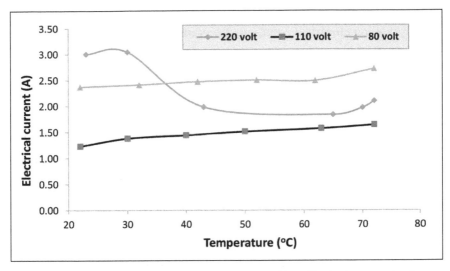

**FIGURE 1.6**  Electrical current passing in milk treated by milk ohmic pasteurizer.[5]

### 1.6.3  MILK TEMPERATURE

Figure 1.7 shows that milk temperature ($T$) was increased with increasing heating time ($t$). The required heating time for temperature to reach 72°C was 6.65, 6.15, 5.15, and 7.65 min for ohmic heating at 80, 110, 220 V,

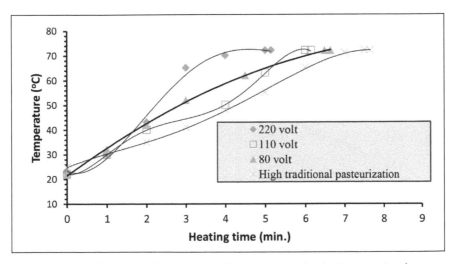

**FIGURE 1.7**  Effects of heating time on milk temperature for ohmic pasteurizer.[5]

and high traditional pasteurization, respectively.[5] Investigators developed following empirical equations for prediction of milk temperature for ohmic pasteurizer:

$$T_{80V} = -0.6246t^2 + 11.799t + 21.478 \qquad (1.36)$$

$$T_{220V} = 0.2507t^4 - 36.25t^3 + 15.095t^2 - 6.4339t + 23.288 \qquad (1.37)$$

$$T_{Cont} = -0.0405t^4 + 0.5209t^3 - 1.6774t^2 - 6.6737t + 24.862 \qquad (1.38)$$

$$T_{110V} = -0.122t^5 + 1.7925t^4 - 8.9003t^3 + 16.887t^2 - 15.958X + 21.989 (1.39)$$

### 1.6.4   THERMAL CONDUCTIVITY

Thermal conductivity of most high moisture foods is nearly similar to that of water. It is an important property that determines thermal transfer through the food during processing and moisture content of food has large impact on the thermal conductivity.[65] Figure 1.8 illustrates the relationship between milk thermal conductivity and temperature for ohmic heating at 80, 110, 220 V, and high traditional pasteurization.[5] Thermal conductivity was increased with increasing temperature at all voltages and high traditional pasteurization. For example, when milk temperatures were 22, 32,

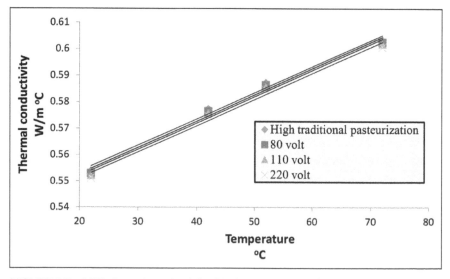

**FIGURE 1.8**   Milk thermal conductivity for ohmic pasteurizer.[5]

42, 52, 62, and 72°C, thermal conductivity was 0.55, 0.56, 0.57, 0.58, 0.59, and 0.6 (W/m$^2$.°C), respectively, for ohmic heating at 80 V. The following empirical equations describe the linear relationships between thermal conductivity and temperature:

$$K_{80V} = 0.001\ T + 0.532 \tag{1.40}$$

$$K_{110V} = 0.001\ T + 0.532 \tag{1.41}$$

$$K_{220V} = 0.001\ T + 0.531 \tag{1.42}$$

$$K_{cont} = 0.001\ T + 0.534 \tag{1.43}$$

## 1.6.5  MILK VISCOSITY

Viscosity is the resistance that liquid offers to movement or flow of its one layer relative to the other. It can also be defined as ratio of cutting resistance to the cutting velocity. It is considered as a tool to monitor the quality of foods at different stage of production. Therefore, it is related to the ability to manipulate the amount of heat used by measuring viscosity, as there is strong relationship between viscosity and heat diffusion. Increase of viscosity hinders the movement of heated membrane and resulted in slowing the heat transfer rate in comparison with low viscosity liquids. Pasteurization at 72°C for 15 s had no effect on milk viscosity.[27] Milk viscosity ranged from 1.4 to 2.2 centipoise at 20°C, as it depends on the milk colloidal particles especially proteins and fats. Viscosity is important milk property to control fatty products.[6]

Figure 1.9 shows that milk viscosity was decreased with increasing temperature for ohmic heating at 80, 110, and 220 V and high traditional pasteurization.[5] At 22°C, milk viscosity was $1.75 \times 10^{-3}$, $1.71 \times 10^{-3}$, $1.37 \times 10^{-3}$, and $1.73 \times 10^{-3}$ pa.s under ohmic heating at 80, 110, and 220 V and high traditional pasteurization, respectively; and it was $2.8 \times 10^{-3}$, $2.39 \times 10^{-4}$, $7.84 \times 10^{-4}$, and $2.8 \times 10^{-4}$ pa.s, respectively, at temperature of 72°C. Milk viscosity was decreased because of reducing fat clusters with increasing temperature. Fat clusters affect the milk viscosity at a particular temperature. Denaturation of whey proteins is sensitive to high temperatures. Also, there is a damage of Gluten proteins, which are responsible for the collection of fat globules. All these factors led to decrease in viscosity. The following empirical equations describe the relationship between milk viscosity and temperature:

$$\mu_{220V} = -2.756 \times 10^{-5}\, T + 0.002 \tag{1.44}$$

$$\mu_{80V} = -2.85 \times 10^{-5}\, T + 0.002 \tag{1.45}$$

$$\mu_{Cont} = -2.923 \times 10^{-5}\, T + 0.002 \tag{1.46}$$

$$\mu_{110V} = 4.312 \times 10^{-7}\, T^2 - 6.559 \times 10^{-5}\, T + 0.003 \tag{1.47}$$

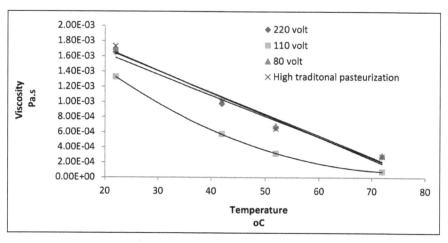

**FIGURE 1.9**   Effects of temperature on milk viscosity under ohmic pasteurizer.[5]

## 1.6.6 MILK DENSITY

Density is a mass of the food per unit volume. The specific gravity of milk is the ratio of milk density to the density of water and can be indicator of adulteration of the milk[8] for cattle milk, specific gravity ranges from 1.035 to 1.051. However, specific gravity of mixed milk ranges from 1.030 to 1.035. The density and temperature are negatively related for most milk as compared to water due to the presence of protein and fat in milk. However, density of lactose is not affected significantly.[8]

Milk specific gravity is the weightage average of specific gravities of its contents. Since the contents are heavily influenced by many factors. The low value of milk specific gravity indicates the addition of water. Measurements are done at temperature of 15.5°C.[6]

Figure 1.10 shows that milk density was decreased with increasing temperature for all ohmic heating treatments.[5] Milk density treated by ohmic heating at 80 V and temperature of 22°C was 1.020 kg/m³. Milk density depends on the content of proteins and fat in milk. If milk fat is reduced, then the density will increase. When milk was pasteurized by ohmic heating at 110 and 220 V, the density was lower than the milk pasteurized by ohmic heating at 80 V and high traditional pasteurization because of heating intensity, which reduces denaturation of whey proteins on the electrodes and led to loss of part of milk proteins. The following empirical equations can be used for the prediction of milk density at different temperatures.

$$\rho_{220V} = -0.308T + 1014.395 \qquad (1.48)$$

$$\rho_{110V} = -0.304T + 1016.674 \qquad (1.49)$$

$$\rho_{220V} = -0.343T + 1028.114 \qquad (1.50)$$

$$\rho_{cont} = -0.303T + 1023.573 \qquad (1.51)$$

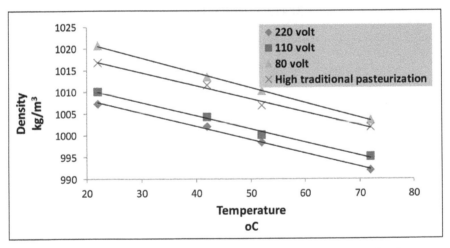

**FIGURE 1.10**  Effects of temperature on milk density under ohmic pasteurizer.[5]

## 1.6.7  SPECIFIC HEAT

It is the quantity of heat energy to rise water temperature by one degree at a constant pressure. It can also be defined as the lost or gained heat of a unit weight of product to reach the preferred temperature without changing the state. It is an important part of thermal analysis in food processing or in food manufacturing. Milk specific heat is influenced by moisture content, temperature, and pressure of the milk.[7, 22] As specific heat increases with rise of product moisture and gas specific heat at constant pressure is greater than that at constant volume, most engineering applications for food processing use specific heat at constant pressure.[24,83]

Figure 1.11 shows that specific heat of milk is increased with increase in temperature for all ohmic heating treatments.[5] For example, at milk temperature of 22 and 72°C, the specific heat was 3860 and 3890 J/kg.°C, respectively, at ohmic heating of 80 V; however, for traditional pasteurization it reached 3870 and 3900 J/kg.°C, respectively. The following empirical equations indicate the linear relationships between the specific heat and temperature:

$$C_{p\,220V} = 3835.752 + 0.611\ T \tag{1.52}$$

$$C_{p\,110V} = 3841.651 + 0.611\ T \tag{1.53}$$

$$C_{p\,80V} = 3847.316 + 0.609\ T \tag{1.54}$$

$$C_{p\,cont} = 3856.206 + 0.605\ T \tag{1.55}$$

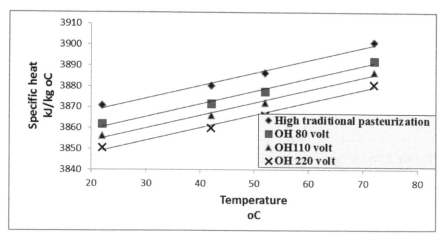

**FIGURE 1.11**    Effects of temperature on milk specific heat for ohmic pasteurizer.[5]

## 1.6.8  THERMAL DIFFUSIVITY

Thermal diffusivity represents the thermal conductivity of food divided by its density and specific heat.[83] Thermal diffusion value of food products ranges from $1 \times 10^{-7}$ to $2 \times 10^{-7}$ m²/s.[83] It can be calculated from the following equation:

$$\propto = \frac{K}{C_p} \qquad (1.56)$$

where $\propto$ = the thermal diffusivity [m²/s], $k$ = the thermal conductivity (W/m².°C), and $C_p$ = the specific heat at constant pressure (J/kg.°C).

There were no significant differences in thermal diffusivity among ohmic heating treatments and high traditional pasteurization (Fig. 1.12). Thermal diffusivity of pasteurized milk by ohmic heating at 80, 110, 220 V, and high traditional pasteurization was $1.47 \times 10^{-7}$, $1.48 \times 10^{-7}$, $1.483 \times 10^{-7}$, and $1.47 \times 10^{-7}$ m²/s, respectively.[5] Thermal diffusivity was increased with increase in temperature because of the thermal diffusivity is directly proportional to extrusive thermal conductivity and inversely related with the density and specific heat. Thermal diffusivity can be calculated by the following linear equations:

$$\lambda_{220V} = 2.787 \times 10^{-10}\, T + 1.365 \times 10^{-7} \qquad (1.57)$$

$$\lambda_{110V} = 2.779 \times 10^{-10}\, T + 1.362 \times 10^{-7} \qquad (1.58)$$

$$\lambda_{80V} = 2.803 \times 10^{-10}\, T + 1.346 \times 10^{-7} \qquad (1.59)$$

$$\lambda_{cont} = 2.736 \times 10^{-10}\, T + 1.352 \times 10^{-7} \qquad (1.60)$$

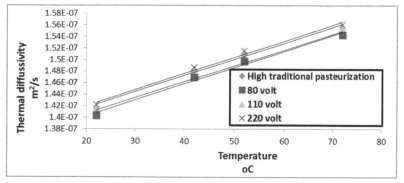

**FIGURE 1.12**   Milk thermal diffusivity versus temperature for ohmic pasteurizer.[5]

## 1.7   PERFORMANCE COEFFICIENT OF THE PROCESSING SYSTEM

System performance coefficient (SPC) is calculated as follows:[35]

$$SPC, = \frac{Q_t}{E_g} \quad \text{where} \tag{1.61}$$

$$E_g = Q_t + E_{\text{loss}} = \sum \Delta V I t \tag{1.62}$$

$$Q_t = m c_p \left( T_f - T_i \right) \tag{1.63}$$

where $m$ = mass (kg), $T_f$= final temperature (°C), $T_i$= primary temperature (°C), $E_g$ = amount of given energy (J) and $Q_t$ = the amount of heat taken (J); $V$ = voltage (V), $I$ = current (A), $t$ = time (s), $c_p$ = specific heat (J/kg. °C), and $E_{\text{loss}}$ = lost heat (J).

It can be observed in Figure 1.13 that the performance factor was decreased with increase in applied voltage. Performance factor reached 0.82 at 80 V then was reduced to 0.7 at 110 V and 0.59 at 220 V because the loss of energy at 220 V was higher than that at 110 and 80 V. Icier and Ilicali[35] and Hosain et al.[33] stated that increasing applied voltage led to reduction in performance factor. They explained that 0–15% of electric energy provided to ohmic heating system was not used for processing of peach juice and it was considered as loss in energy. They stated that increasing performance factor refers to the reduction in energy loss and most of the electric energy was converted to heat energy at zero energy loss. The maximum value of

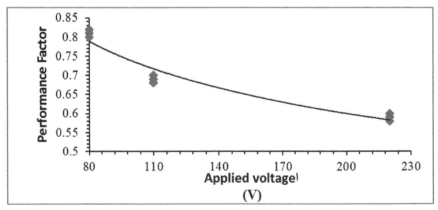

**FIGURE 1.13**   Performance coefficient for milk ohmic pasteurizer versus applied voltage.[5]

performance factor is 1 or 100%. The relationship between performance factor and applied voltage is shown in Figure 1.13:[13]

$$\eta = 2.9264 \ V^{-0.299} \qquad (1.64)$$

### 1.7.1 HEATING RATE

Heating rate was increased with increase in applied voltage. Increasing voltage means increasing power supplied to the ohmic heating system. The heating rate was 10.9, 11.61, and 13.84°C/min at applied voltage of 80, 110, and 220 V, respectively (Fig. 1.14). The relationship between heating rate and voltage was linear with a coefficient determination of 0.991 as indicated in the following equation:

$$0.022 \ V + 9.162 = H_R \qquad (1.65)$$

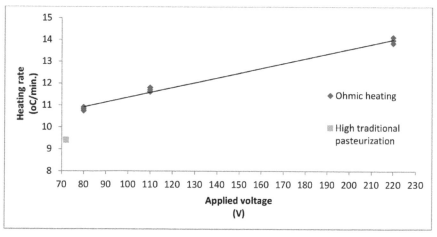

**FIGURE 1.14**   Milk heating rate versus applied voltage for milk ohmic pasteurizer.[5]

## 1.8   HOLDING TIME OF MILK IN THE OHMIC PASTEURIZER

Holding time of milk for the ohmic pasteurizer is calculated as follows:

$$t = \frac{\rho A L_d}{W} \qquad (1.66)$$

where $t$ = the holding time (s), $\rho$ = density (kg/m$^3$), $A$ = the area (m$^2$), $L_d$ = the tube length (m), and W = mass flow (kg/s).

Figure 1.15 shows the relationship between milk temperature and holding time using ohmic heating at different voltages (80, 110, and 220V). During milk pasteurization by ohmic heating, milk is held every part of pasteurizer for a specific time that is different for each subprocess. The required time for raising milk temperature from 22 to 40°C is 3.96 min for the heat exchanger, and then milk is brought to ohmic heating unit and stays there.[5] The required time for reaching milk temperature to 72°C is 1.51 min; and after this milk temperature is reduced to 22°C for a holding time of 0.34 min. Therefore, the total holding time for the ohmic pasteurizer at 80 V is 6.3 min. However, the required time for high traditional pasteurization and ohmic pasteurizing at 80, 110, and 220 V was 11.11, 6.3, 6, and 5.54 min, respectively.

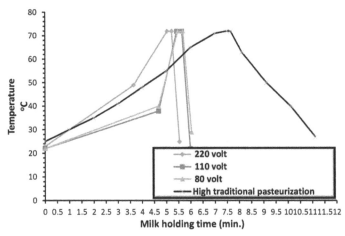

**FIGURE 1.15** Milk temperatures (heating and cooling) versus holding time for ohmic pasteurizer.[5]

## 1.9 EFFECT OF OHMIC HEATING ON THE CHEMICAL COMPOSITION OF MILK

### 1.9.1 MILK PROTEINS

The quantity of milk deposits depends on type of raw milk, type and intensity of heat treatment. It has been found that deposits increase with heat treatment intensity in any thermal treatments. It is believed that percentage of deposits of milk solids is the residual part of solids that is deposited on the surface of heat exchanger, thus causing fouling. Sandu and Lund[73] have discussed theoretical mechanism of milk deposits in details. They suggested the presence of at least two separate interactions in heat-treated milk that contribute to the formation of deposits.

- The first is whey protein denaturation, responsible for smooth protein deposits (smaller particles). Whey protein denaturation plays an important role in the formation of deposits on exchanger surfaces during milk and its products during manufacturing.
- The second is a decrease of calcium triphosphate that is dissolved with rising temperature, thus resulting in bigger particles.

Rakes et al.[64] found that formation of deposits within thermal exchanger because whey proteins are much higher, during exposure of milk to the thermal treatment. In dairy factories, deposits are formed at different apparatuses; pasteurization devices or evaporators. Cleaning of these devices is a costly operation, and requires high energy, cleaning materials, and labor.[54]

Deposits deteriorate the quality of product, because milk cannot be heated to a required temperature during pasteurization.[13,82] These deposits reduce heat transfer efficiency and increase the downward pressure.[82] Johansson[39] indicated that deposits are part of the problems that occur during the use of HTST method in pasteurization as proteins and minerals are deposited on thermal exchanger surfaces. However, Ayadi et al.[10] and Pereira et al.[61] showed that ohmic heating improves product quality by reducing deposits. Also, hot surfaces in ohmic heating are minimum, which is considered as an important factor in reducing problems of deposits.[28]

Huixian et al.[34] suggested that there is no significant difference between ohmic heating and traditional method as whey protein is denatured at all temperatures. However, Pereira et al.[62] found significant differences between ohmic heating and traditional method in their effects on whey protein denaturation. Ohmic heating is less denaturant due to absence of hot surfaces and volume heating of milk. Milk temperature has profound effect on proteins. Sensitivity of proteins to temperature varies. For example, casein proteins bear high temperatures higher than whey proteins which are affected by pasteurization temperature. However, casein proteins are affected by heat treatment at temperature higher than 100°C. The results showed that the protein percentage in pasteurized milk by ohmic heating at 110 and 220 V was reduced from 3.6% in the raw milk to 3.5 and 3.54%, respectively (Table 1.1). This is because of changing some whey proteins at 27°C temperature and 5–10% of deposits, while deposits were 3.57 and 3.58% at 80 V and high traditional pasteurization, respectively.

## 1.9.2 MILK FATS

Heat treatment does not have a pronounced effect on the milk fat percentage, but it affects the vitamins soluble in fats. In fact, effect of pasteurization temperature on milk fat is very low. Table 1.1 shows that milk fat percentage was reduced from 3.7% in the raw milk to 3.6 in pasteurized milk by ohmic heating at 220, 110, and 80 V, respectively, while it did not change in milk treated by high traditional pasteurization (3.7%) because of clustering of some fat globules in the holding tube.[5]

**TABLE 1.1**   Chemical Composition of Milk Treated by Ohmic Pasteurizer.[5]

| Chemical composition | Before pasteurization % | High traditional pasteurization % | Ohmic heating at | | |
|---|---|---|---|---|---|
| | | | 220 V | 110 V | 80 V |
| Ash | 0.68 | 0.71 | 0.73 | 0.69 | 0.69 |
| Protein | 3.6 | 3.57 | 3.5 | 3.54 | 3.58 |
| Fat | 3.7 | 3.7 | 3.6 | 3.6 | 3.6 |
| Lactose | 5.02 | 6.0 | 7.1 | 6.9 | 6.2 |
| Moisture content | 87 | 86.0 | 85.0 | 85.2 | 85.9 |

## 1.9.3 MILK LACTOSE

It can be seen from Table 1.1 that lactose percentage after pasteurization was changed significantly. It was 5.02% before pasteurization and then was increased to 6, 6.9, 7.1, and 6% after pasteurization by using ohmic heating at 80, 110, 220 V, and high traditional pasteurization, respectively.[5]

## 1.9.4 MILK ASH

Ash percentage in milk is affected with sterilization or pasteurization heat (Table 1.1). Ash percentage in raw milk was 0.68% then was increased to

0.73% in the pasteurized milk by ohmic heating at 220 V and it was 0.69% in pasteurized milk by ohmic heating both at 110 and 80 V. However, ash percentage in pasteurized milk by high traditional pasteurization method was 0.71% because the citrates were burnt; also $CO_2$ and carbonates were evaporated during the incineration.

### 1.9.5  MOISTURE CONTENT OF MILK

It can be seen in Table 1.1 that milk moisture content of raw milk was 87% compared to 85, 85.2, 85.9, and 86% in pasteurized milk at 220, 80, 110 V, and high traditional pasteurization, respectively. This is due to differences between raw milk temperature and temperature of pasteurized milk by ohmic heating.

## 1.10  EFFECTS OF OHMIC HEATING ON PHYSICAL PROPERTIES OF MILK

### 1.10.1  MILK ACIDITY

The acidity of raw milk ranges from 0.14 to 0.18%. Milk acidity is due to the presence of milk proteins, $CO_2$, and some acid salts, which are present in raw milk naturally. Table 1.2 illustrates that raw milk acidity is 0.15%, which is within the permitted range; however, the acidity was reduced to 0.14, 0.14, 0.13, and 0.14 % in pasteurized milk by ohmic heating at 220, 110, 80 V, and high traditional pasteurization, respectively, due to loss of soluble gases like $CO_2$ lost and conversion of soluble calcium phosphate ratio into glutinous state.[5]

**TABLE 1.2**  Physiochemical Characteristics of Milk Treated by Milk Ohmic Pasteurizer.[5]

| Physiochemical characteristic | Before pasteurization | High traditional pasteurization | Ohmic heating | | |
|---|---|---|---|---|---|
| | | | 220 V | 110 V | 80 V |
| Total acidity | 0.15 | 0.14 | 0.15 | 0.14 | 0.14 |
| pH | 6.6 | 6.8 | 6.6 | 6.7 | 6.8 |
| Alkaline phosphatase | + | − | − | − | − |

## 1.10.2   pH OF MILK

pH is the negative logarithm of hydrogen ion in the media. Acid particles produce hydrogen ion. The pH is an inverse of the acidity because the source of acidity is lactic acid. The pH measurement refers to the concentration of hydrogen ions, which are produced from ionization of these compounds in acidic milk. The pH of raw milk is 6.6 and was reduced to 6.7, 6.8, 6.8, and 6.8 in pasteurized milk by ohmic heating at 220, 110, 80 V, and high traditional pasteurization, respectively. This increase in pH is due to decline in milk acidity.

## 1.11   DETECTION OF ACTIVITY OF ALKALINE PHOSPHATASE IN MILK

Phosphatase is one of the important enzymes that must be tested to find pasteurization efficiency. Phosphatase is affected by pH and increase in temperature. The results showed that phosphatase is present (positive sign) in raw milk (Table 1.2), while it was absent in pasteurized milk for all treatments.

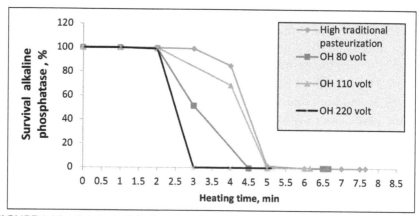

**FIGURE 1.16**   Survival of alkaline phosphatase in milk treated by milk ohmic pasteurizer (OH).[5]

The required time for elimination of phosphatase was 5, 6.5, 7, and 7.5 min for pasteurized milk by ohmic heating at 220, 110, 80 V, and high traditional pasteurization, respectively. These results indicate that the heating, electrical field and pressure (1.25–1.5 bar) using ohmic heating at 220, 110, and 80 V caused inactivation of alkaline phosphatase faster than high traditional pasteurization. Moreover, the survival alkaline phosphatase ratio in pasteurized milk was dropped to zero (Fig. 1.16). The survival alkaline phosphatase ratio is calculated from the following equation.[50]

$$\log\left(\frac{N_j}{N_{oj}}\right) = -\int_o^t \frac{dt}{Dj} \tag{1.67}$$

$$D_j = D_{rj} 10^{\frac{T_{ri}-T}{Z}} \tag{1.68}$$

where $N_j / N_{oj}, t, D_j, D_{rj}, T_{ri}, T,$ and $z$ are ratio of survival phosphatase, time (min), the decimal reduction time to lethal of 90% of phosphatase during one logarithmic cycle at specific temperature, the decimal reduction time to lethal of 90% of phosphatase during one logarithmic cycle at another temperature, pasteurization temperature, temperature, thermal resistance (°C), respectively; and $D_{rj}$ is calculated from the following equation:[90]

$$D_{rj} = \frac{(t_2 - t_1)}{logN_{oj} - log\ N_j} \tag{1.69}$$

where $N_{oj}, N_j$ are count of initial microorganisms and the final microorganisms, respectively.

Thermal resistance ($z$) is calculated from the following equation:[90]

$$Z_j = \frac{(T_2 - T_1)}{\log D_{rj} - \log D_j} \tag{1.70}$$

where $T_2 - T_1$ is the difference of temperature.

Rao et al.[65] found that values of the $D$ and $Z$ for phosphatase were 24.6 min and 5.4°C, respectively. These equations can be used to calculate the ratio of survival microorganism.

## 1.12 EFFECTS OF OHMIC HEATING ON ELIMINATION OF MICROBES

Ohmic heating technique has an important role in the elimination of live microbes that are present in raw milk. Many studies indicated the ability of ohmic technique in the elimination of microbes due to deadly effect of thermal and non-thermal (rotary current) methods, especially viable aerobes and *Streptococcus thermophilus* 2646. Also, many studies showed the important role of rotary current to get rid of harmful microbes in milk.[57,76–80] The researchers also indicated that the inhibition effect of rotatory current depends on electric power, the current going through the media and the time for which the cells leave the recession in the media after electrical treatment. Palaniappen et al.[55,56] found that there were no differences between ohmic heating and traditional method using same temperature in eliminating yeasts at electrical current of 0.5–1 A at 60 Hz.

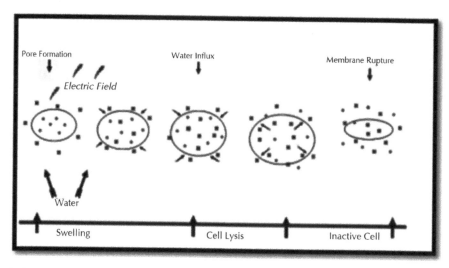

**FIGURE 1.17**    Stages of electroporation in the cell membrane.

Huixian et al.[34] found significant differences between pasteurization by ohmic heating and traditional method in microbes death especially psychotropic. They revealed the importance of electricity in elimination rate in comparison with heating. Therefore, it is obviously clear that non-heating mechanical (electricity) of ohmic technique is effective to eliminate microbes.[59,63] Figure 1.17 illustrates the stages of electroporation in the cell membrane.[5]

## 1.13   EFFECTS OF PASTEURIZATION TEMPERATURE ON MILK BACTERIOLOGICAL, CHEMICAL, AND ORGANOLEPTIC CHARACTERISTICS

The objective of heat treatment is to inhibit pathogenic microbes and to prolong period of milk preservation. This treatment can affect milk properties and milk components, such as the effect on apparent sensory characters, as pasteurization temperature does not affect milk color and taste due to the expulsion of many volatile and odd contents.[8] Milk microbial content is influenced by pasteurization temperature, which inhibits a large number of microbes such as psychrophilic bacteria (endemic in milk), all pathogenic as *Escherichia coli* (an indicator of the efficiency of the pasteurization and its existence proof a lack of a properly pasteurization), mold and yeast.[8]

Razzak et al.[67] found a decrease in bacterial total count from 5.01 × $10^4$ cfu/mL to 3.1 × $10^4$ cfu/mL when they used solar pipe heating system for milk pasteurization. They also found a decrease in the total count of coli from 4.2 × $10^4$ to zero. Milk Standards in Australia[85] indicated that total bacteria count in pasteurized milk is around 50 × $10^3$–100× $10^3$ cfu/mL and that of *E. coli* is 10 cfu/mL. ICMSF[37] clarified that pathogens such as *Campylobacter, Salmonella, E. coli, Staphylococcus, Yersinia enterocolitica*, and *Listeria monocytogenes* were totally eliminated by the pasteurization temperature. Microbes with endospores such as *Bacillus* and *Paenibacillus* can be totally eliminated by using HTST method. Milk protein can be divided into two types: casein proteins representing main part of milk protein are not affected by pasteurization temperature; whey proteins change their attributes when exposed to pasteurization temperature causing 5–10% of deposits. This change may obstacle some processing methods and does not decrease the milk nutritional value; however, nitrogen distribution is affected in a limited way in comparison with sterilization.[8] Pasteurization causes whey protein denaturant, which does not influence the milk nutritional value because it affects some bands only that are responsible for protein stability and not peptide bands among amino acids.[27] Also, pasteurization neither affects the nutritional value of milk fat but a carbonyl compounds resulted from the effect of heat have an impact on milk flavor, nor unsaturated or essential fatty acids.[27] Fat particles are influenced by pasteurization, as heating of milk changes its size than in raw milk.[20] Walstren et al.[95] indicated that pasteurization caused little changes in particles membrane. Lactose is not affected by pasteurization neither quantity nor quality except raining of temperature higher than the pasteurization temperature.[8]

Milk dissolved calcium and phosphorus are decreased on heating of milk. This decrease depends on temperature of exposure. However, pasteurization has low effect on milk minerals content and does not affect its nutritional value.[27] Large percent of vitamins (vitamin dissolved in fat are A, D, E, and K) are not sensitive to thermal treatment of pasteurization, therefore no differences in milk vitamin contents were observed. Whereas, vitamins dissolved in water are sensitive to pasteurization thermal treatments.[27] Pasteurization leads to loss of dissolved gases in milk especially $CO_2$, which is most important factor for acidity reaction of milk resulting in relative increase in pH and decrease in acidity.[8] However, Walstren et al.[95] found that pH of pasteurized milk were lower than that of raw milk. Tamime[88] found that the pH of raw milk was 6.6 and it was decreased to 6.3 when pasteurization was done at 72°C for 15 s. Endemic enzyme in milk was influenced by pasteurization temperature especially alkaline phosphatase, which is the formal test for pasteurization accuracy and indicator of any error during pasteurization or adding very low amount (not more than 0.2%) of raw milk to pasteurized milk.[49]

In case of incomplete inhibition of enzyme, proteins, and fats will hydrolyze, which can have negative effects on quality of milk and its products.[93] Fenoll et al.[26] stated that alkaline phosphatase was degraded by pasteurization temperature and it exists only in raw milk. Pasteurization of milk and products could last for 48 h without spoilage at 20°C (no cooling) and for 10–16 days at 4°C by using good quality milk and using best hygiene practices. This can be done by using fast or slow pasteurization. In a case study in Australia, the best temperature was 72–78°C for 15–25 s, and ideal environment of pasteurization varies with its type and country.[40]

### 1.13.1 TOTAL COUNT OF BACTRIA

Total count of bacteria were $114 \times 10^3$, 0, 0, 0, and 100 CFU/mL in raw milk, pasteurized milk by ohmic heating at 220, 110, 80 V, and high traditional pasteurization, respectively. These values are lower than the minimum limit of standard specification in Australia which is between 50,000–100,000 CFU/mL of total count of Bactria in pasteurized milk. Figure 1.18 shows that the ratio of survival microorganisms (%) was reduced with increasing time because of increase in electric current passing through the milk and heating.[34] The required time for elimination of microorganisms was 5, 6, and 6.5 min for ohmic heating at 80, 110, and 220 V, respectively.

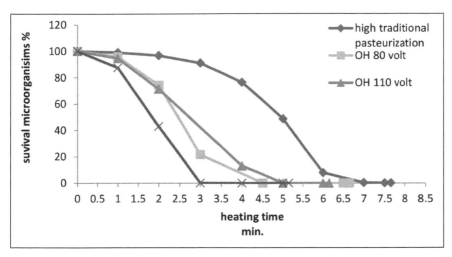

**FIGURE 1.18** Survival microorganism's ratio (%) in milk treated by ohmic pasteurizer.[5]

## 1.13.2 ESCHERICHIA COLI

Table 1.3 shows that *E. coli* in raw milk was 99,000 CFU/mL and was eliminated completely in pasteurized milk for all treatments.[5] Ratio of survival *E. coli* bacteria reached zero after 5, 6, 6.6, and 7 min by using ohmic heating at 220, 110, 80 V, and high traditional pasteurization, respectively (Fig. 1.19). Also, Figure 1.20 shows that staph-110, yeasts, and mold were absent in the pasteurized milk. Ratio of survival staph-110 reached zero for all treatments.

**TABLE 1.3** Microorganism Count in the Milk Treated by Ohmic Pasteurizer.[5]

| Microorganisms | Before pasteurization | High traditional pasteurization | Ohmic heating | | |
|---|---|---|---|---|---|
| | | | 220 V | 110 V | 80 V |
| Total count | $114 \times 10^3$ | $1 \times 10^1$ | – | – | – |
| *E. coli* | $99 \times 10^3$ | – | – | – | – |
| Staph-110 | $98 \times 10^3$ | – | – | – | – |
| Yeasts | $0.01 \times 10^3$ | – | – | – | – |
| Mold | $0.01 \times 10^3$ | – | – | – | – |

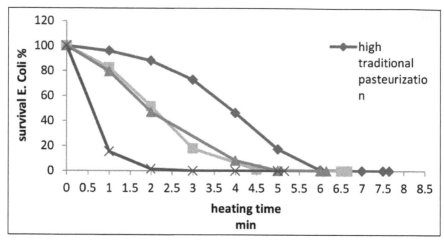

**FIGURE 1.19**   Survival *E. coli* ratio (%) in milk treated by ohmic pasteurizer.[5]

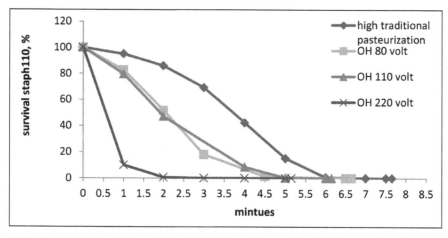

**FIGURE 1.20**   Survival staph-10 ratio (%) in milk treated by ohmic pasteurizer.[5]

### 1.13.3   LETHALITY OF MICROORGANISMS IN MILK

Lethality is a general method to evaluate the heat treatment. Time of heating for lethal of microbes is specified for manufacturing process and is considered for all temperatures. Also the area under curve indicates the effect of heat death per unit time. Lethality is calculated from the following equation:[32]

$$F_{Tref} = \int_o^t L_t(t)\,dt = \int_o^t 10^{\frac{T_{(t)}-T_{ref}}{z}}\,dt \qquad (1.71)$$

Figure 1.21 illustrates that the mean lethality of total heat process represented by calculating the area under curve was 0.766, 0.783, 0,991, and 0.611 min for ohmic heating at 220, 110, 80 V, and high traditional pasteurization, respectively.[5] Also, the lethality in the holding tube was 0.25 min for all treatments. These results indicate the efficiency of heat treatment and are considered good because phosphatase was eliminated completely during 0.25 min.

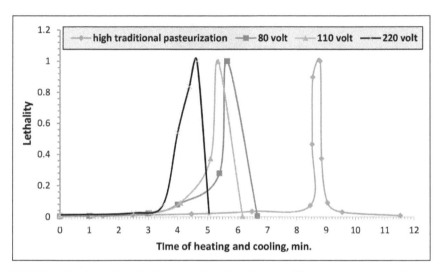

**FIGURE 1.21**    Lethality of microorganisms in milk treated by ohmic pasteurizer.[5]

## 1.14   EFFECT OF OHMIC HEATING ON DENATURATION OF MILK PROTEINS

It can be seen in Figure 1.22 that whey proteins were affected by ohmic heating at 220 V more than that at 110 and 80 V because the increasing voltage causes maximum denaturation of whey proteins. Whey proteins were denatured in pasteurized milk by using ohmic heating at 80 V because of the low voltage of pasteurization.[29] Pereira et al.[62] stated that increasing milk temperature leads to higher denaturation of whey proteins.

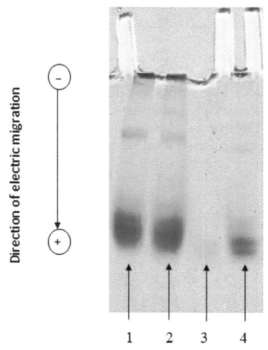

**FIGURE 1.22**    Electric migrations of whey proteins in milk treated by ohmic pasteurizer.[5]

## 1.15   FUTURE PROSPECTIVES AND RESEARCH OPPORTUNITIES

Ohmic heating is a novel technology and it needs deep studies for development of ohmic pasteurizers. Invention of new electrodes is necessary for reducing fouling on surfaces of the electrodes. Also, these should be manufactured for large capacity ohmic pasteurizers, which should be able to operate at low voltages.

## 1.16   CONCLUSIONS

Milk can be pasteurized by ohmic heating at 80 V only. High voltage is unsuitable for milk pasteurization. Chemical, physical, and microbiological properties of pasteurized milk by using ohmic heating at 80 V were better than that at 110 and 220 V. Whey proteins were affected by ohmic heating at 220 V more than that at 110 and 80 V.

## KEYWORDS

- batch pasteurization
- denaturation
- energy efficiency
- ohmic heating technique
- ohmic milk pasteurizer
- ohmic pasteurizer
- pasteurization efficiency
- performance coefficient
- thermal diffusion
- vacuums pasteurization
- *Yersinia enterocolitica*

## REFERENCES

1. Al-Rubaiy, H. H. M. Designing and Manufacturing Milk Pasteurization Device by Solar Energy and Studying Its Efficiency. Ph.D. Thesis, University of Basrah, Iraq, 2010; p 167.
2. Al-Dehan, A. *Food Engineering and Dairy;* Sima-rotomacah: Torcy, France, 1981; p 310.
3. Al-Hilphy, A. R. S. Drying of Corn by Solar Energy. *Basrah J. Agric. Sci.* **2010,** *32* (1), 255–264.
4. Al-Hilphy, A. R. S.; Al-Timimi, A. B. R.; Al-Seraih, A. Milk Pasteurization by Microwave and Study its Chemical and Microbiological Characteristics during Different Storage Times. *J. Basrah Res. (Sciences).* **2010,** *36* (15), 66–76.
5. Al-Hilphy, A. R. S.; Haider, I. A.; Mohsin, G. F. Designing and Manufacturing Milk Pasteurization Device by Ohmic Heating and Studying its Efficiency. *J. Basrah Res. (Sciences): Acad. Sci. J.* **2012,** *4* (38), 1–18.
6. Al-Nimer, T. M. *Theoretical Dairy and Application;* A Knowledge Bestan for Printing, Publishing and Distribution: Alexandria University, Egypt, 2003; p 224.
7. Al-Sharifi, H. R.; Muhammed, S. H. *Practical Dairy Microbiology,* 1st ed.; Dar Alhikmah Press: Basrah, Iraq, 1992; p 784.
8. Al-Shebiby, M. M. A.; Shukri, N. A; Tuamah, S. J.; Ali, H. G. *Dairy Fundamentals;* Dar Alhikma for Press and Publishing: Musol, Iraq, 1980; p 150.
9. Alvin, M. H.; Erlbach, E. *Schaum's Outline of Theory and Problems of Beginning Physics II;* McGraw-Hill Professional: New York, 1998; p 140.

10. Ayadi, M.; Leulieta, J.; Chopard, F.; Berthou, M.; Lebouche, M. Engineering and Chemical Factors Associated with Fouling and Cleaning in Milk Processing. *Innov. Food Sci. Emerg. Technol.* **2004,** *5,* 465–473.

11. Ayadi, M. A.; Bouvier, L.; Chopard, F.; Berthou, M.; Leuliet, J. C. Heat Treatment Improvement of Dairy Products via Ohmic Heating Processes: Thermal and Hydro-dynamic Elect on Fouling. In *Heat Exchanger Fouling and Cleaning: Fundamentals and Applications;* ECI Digital Archives Pub., 2004. http://dc.engconfintl.org/heatex-changer/19/(accessed on May 25, 2016).

12. Baigrie, B. *Electricity and Magnetism: A Historical Perspective*; Greenwood Press: Westport, 2006; p 7–8.

13. Bansal, B.; Chen, X. Critical Review of Milk Fouling in Heat Exchanger. *Compr. Rev. Food Sci. Food Saf.* **2006,** *5* (2), 27–33.

14. Berk, Z. Food Process Engineering and Technology. *A Volume in Food Science and Technology*; Academic Press: Cambridge, MA, 2009; p 230.

15. Berthou, M.; Chopard, F.; Aussudre, C. Heater for Resistive Heating of a Fluid, Fluid-Treatment Apparatus Incorporating Such a Heater, and a Method of Treating a Fluid by Resistive Heating. Word Patent No 0141509, 2001; p 80.

16. Brown, C. E. *Power Resources*; Springer: New York, 2002; p 230.

17. Carlisle, R. *Scientific American Inventions and Discoveries;* John Wiley & Sons Inc.; Hoboken, NJ, 2004; p 357.

18. Castro, I.; Teixeira, J.; Vicente, A. The Influence of Field Strength, Sugar and Solid Content on Electrical Conductivity of Strawberry Products. *J. Food Process Eng.* **2003,** *26,* 17–29.

19. Chalmers, G. K. The Lodestone and the Understanding of Matter in Seventeenth Century England. *Phil. Sci.* **1937,** *4* (1), 75–95.

20. Dalgleish, D. G.; Bank, J. M. Interactions between Milk Serum Proteins and Synthetic Fat Globule Membrane during Heating of Homogenized Whole Milk. *J. Agric. Food Chem.* **1991,** *41* (9), 1007–1012.

21. Davis, J. G. *Dictionary of Drying*, 2nd ed.; Leonard Hill: London, 1955; p 200.

22. De-Alwis, A.; Fryer, P. Finite-element Analysis of Heat Generation and Transfer during Ohmic Heating of Food. *Chem. Eng. Sci.* **1990,** *45* (6), 1547–1559.

23. Earle, R. L. *Unit Operation in Food Processing,* 2nd ed.; Pergamon Press: Oxford, 1992; p 190.

24. Fellows, P. *Food Processing Technology: Principles and Practice;* 2nd ed.; Wood head: Cambridge, England, 2000; p 400.

25. Fenoll, J.; Jourquin, G.; Kauffmann, J. M. Fluorimetric Determination of Alkaline Phosphatase in Solid and Fluid Dairy Products. *Talanta.* **2002,** *56* (6), 1021–1026.

26. FDACFSAN (Food and Drug Administration-Center for Food Safety and Applied Nutrition). *Kinetics of Microbial Inactivation for Alternative Food Processing Tech-nologies-Ohmic and Inductive Heating,* 2000. http://www.fda.gov/Food/FoodScien-ceResearch/SafePracticesforFoodProcesses/ucm101246.htm (accessed on April 15, 2016).

27. Fryer, P. J.; De-Alwis, A. A. P.; Koury, E.; Stapley, A. G. F.; Zhang, L. Ohmic Processing of Solid-liquid Mixtures: Heat Generation and Convection Effects. *J. Food Eng.* **1993,** *18* (2), 101–125.

28. Gaffar, A. M. *Heat Treatments in the Dairy Plants;* Arabic Dar for Publishing and Distribution: Ain Shams University, Egypt, 2001; p 150.

29. Getchel, B. Electric Pasteurization of Milk. *Agric. Eng.* **1935,** *16* (10), 408–410.

30. Ghnimi, S.; Flach-Malaspina, N.; Dresh. M. Evaluation of an Ohmic Heating Unit for Thermal Processing of Highly Viscous Liquids. *Chem. Eng. Res. Des.* **2008,** *86,* 627–632.

31. Gut, J. A. W.; Fernandes, R.; Tahini, C. C.; Pinto, J. M. In *HTST Milk Processing: Evaluating the Thermal Lethality Inside Plate Heat Exchangers*, ICEF9–2004, International Conference on Engineering and Food, 2004. http://sites.poli.usp.br/pqi/lea/docs/icef2004a.pdf (accessed on April 13, 2016).

32. Hosain, D.; Adel, H.; Farzad, N.; Mohammad, H. K.; Hosain, T. Ohmic Processing: Temperature Dependent Electrica Conductivities of Lemon Juice. *Modern Appl. Sci.* **2011,** *5* (1), 209–216.

33. Huixian, S.; Shuso, K.; Jun-ichi, H.; Kazuhiko, I.; Tatsuhiko, W.; Toshinori, K. Effect of Ohmic Heating on Microbial Counts and Denaturation of Protein in Milk. *Food Sci. Technol. Res.* **2008,** *14* (2), 117–123.

34. Icier, F.; Ilicali, C. Effects of Concentration on Electrical Conductivity of Orange Juice Concentrates during Ohmic Heating. *Eur. Food Res. Technol.* **2005,** *220,* 406–414.

35. Icier, F.; Yildiz, H.; Baysal, T. Polyphenoloxidase Deactivation Kinetics during Ohmic Heating of Grape Juice. *J. Food Eng.* **2008,** *85,* 410–417.

36. ICMSF (International Commission on Microbiological Specifications for Foods). *Microorganisms in Food 6: Microbial Ecology of Food Commodities;* Kluwer Academic/Plenum Publishers: New York, 2005; p 233.

37. James, W. N.; Susan, A. *Electric Circuits.* Prentice Hall: Upper Saddle River, NJ, 2008; p 29.

38. Johansson, C. *Heating of Milk and Milk Fouling;* Technical Report: Linkopings Universitet - Projekt i mikrobiell bioteknik *TVMB04,* 2008. www.liu.se/content/1/c6/11/88/57/2008/Cccilia_Johansson.pdf (accessed on April 19, 2016).

39. Juffs, H.; Deeth, H. C. Scientific Evaluation of Pasteurization for Pathogen Reduction in Milk and Milk Products. *Food Standards, Commonwealth of Australia New Zealand*; Kingston ACT 2604, Australia; 2000; p. 110. https://www.foodstandards.gov.au/code/proposals/documents/Scientific%20Evaluation.pdf (accessed on May 27, 2016).

40. Jun, S.; Sastry, S. K. Modeling and Optimization of Pulsed Ohmic Heating of Food Inside the Flexible Package. *J. Food Eng.* **2005,** *28* (4), 417–436.

41. Jun, S.; Sastry, S. K. Reusable Pouch Development for Long Term Space Mission: A 3D Ohmic Model for Verification of Sterilization Efficacy. *J. Food Eng.* **2007,** *80,* 1199–1205.

42. Kemp, M. R.; Fryer, P. J. Enhancement of Diffusion Through Foods Using Alternating Eelectric Fields. *Innov. Food Sci. Emerg. Technol.* **2007,** *8,* 143–153.

43. Kim, H. J.; Choi, Y. M.; Yang, A. P. P.; Yang, T. C. S.; Taub, I. A.; Giles, J.; Ditusa, C.; Chall, S.; Zoltai, P. Microbiological and Chemical Investigation of Ohmic Heating of Particulate Foods Using a 5 kW Ohmic System. *J. Food Process. Preserv.* **1996,** *20* (1), 41–58.

44. Kong, Y. Q.; Dong, L.; Wang, L. J.; Bhandari, B.; Chen, X. D.; Mao, Z. H. Ohmic Heating Behavior of Certain Selected Liquid Food Materials. *Int. J. Food Eng.* **2008,** *4* (3), Article 2. DOI: 10.2202/1556-3758.1378

45. Leizerson, S.; Shimoni, E. Stability and Sensory Shelf Life of Orange Juice Pasteurized by Continuous, Ohmic Heating. *J. Agric. Food Chem.* **2005,** *53,* 4012–2018.

46. Lima, M. Ohmic Heating: Quality Improvements. *Encyclopedia Agric. Food Biol. Eng.* **2007,** *1,* 1–3.

47. Ma, Y.; Borbano, D. M. Milk pH as a Function $CO_2$ Concentration, Temperature and Pressure in a Heat Exchanger. *J. Dairy Sci.* **2003,** *86* (12), 3822–3830.

48. Mahmood, A. A. *Liquid Milk (Practical);* Alhekmah Press for Printing: Basrah, Iraq, 1986; p 200.

49. Maroulis, Z. B.; Saravacos, G. D. *Food Process Design;* Marcel Dekker Inc.: New York, 2003; p 240.

50. Marra, F.; Zell, M., Lyng, J. G.; Morgan, D. J.; Cronin, D. A. Analysis of Heat Transfer during Ohmic Processing of a Solid Food. *J. Food Eng.* **2009,** *91,* 56–63.

51. Moses, B. D. Electric Pasteurization of Milk. *Agric. Eng.* **1938,** *19,* 525–526.

52. Schade, A. L. Prevention of Enzymatic Discoloration of Potatoes. US Patent 2,569,075, 1951; p 16.

53. Noguchi, A. *Potential of Ohmic Heating and High Pressure Cooking for Practical Use in Soy Protein Processing;* Japan International Research Center for Agricultural Sciences, Parker, Thomas Robertson. Patent for method of and line for bread crumb production, 2002; p 50.

54. Novy, M.; Zany, R. *Identification of Fouling Model in Flow of Milk at Direct Ohmic Heating;* CTU in Prague: Prague 6, 2004; p 31.

55. Palaniappan, S.; Sastry, S. K.; Richter, E. R. Effects of Electro Conductive Heat Treatment and Electrical Pretreatment on Thermal Death Kinetics of Selected Microorganisms. *Biotechnol. Bioeng.* **1992,** *39,* 225–232.

56. Palaniappan, S.; Sastry, S. K. Electrical Conductivity of Selected Juices: Influences of Temperature, Solids Content, Applied Voltage, and Particle size. *J. Food Process Eng.* **1991,** *14,* 247–260.

57. Pareilleux, A.; Sicard, N. Lethal Effects of Electric Current on *Escherichia coli. J. Appl. Microbiol.* **1970,** *19,* 421–424.

58. Parrot, D. L. Use of Ohmic Heating for Aseptic Processing of Food Particulates. *Food Technol.* **1992,** *46* (12), 68–72.

59. Paul, S.; Heldman, D. R. *Introduction to Food Engineering;* Academic Press Inc.: San Diego, CA, 1993; p 381.

60. Pereira, R.; Martins, J., Mateus, C.; Teixeira, J. A.; Vicente, A. A. Death Kinetics of *Escherichia coli* in Goat Milk and *Bacillus licheniformis* in Cloudberry Jam Treated by Ohmic Heating. *Chem. Papers.* **2007,** *61* (2), 121–126.

61. Pereira, R. N.; Martins, R. C.; Vicente, A. A. Goat Milk Free Fatty Acid Characterization during Conventional and Ohmic Heating Pasteurization. *J. Dairy Sci.* **2008,** *91,* 2925–2937.

62. Pereira, R. N.; Teixeira, J. A.; Vicente, A. A. *Denaturation of Whey Proteins of Milk during Ohmic Heating.* IBB (Institute for Biotechnology and Bioengineering), Portugal, 2010; p 90.

63. Shirsat, N.; Lyng, J. G.; Brunton, N. P.; McKenna, B. Ohmic Processing: Electrical Conductivities of Pork Cuts. *Meat Sci.* **2004,** *67* (3), 507–514.

64. Rakes, P. A.; Swartzel, K. R.; Jones, V. A. Deposition of Dairy Protein-containing Fluids on Heat Exchanger Surfaces. *Biotechnol. Progr.* **1986,** *2* (4), 210–217.

65. Rao, M. A.; Rizvi, S. S. H.; Datta, A. K. *Engineering Properties of Food;* 3rd ed.; CRC Press: Boca Raton, FL,, 2005; p 732.

66. Raynal-Ljutovac, K.; Park, Y.; Gaucheron, F.; Bouhallab, S. Heat Stability and Enzymatic Modifications of Goat and Sheep Milk. *Small Rumin. Res.* **2007,** *68,* 207–220.

67. Razzak, F. H.; Ayoob, N. Y.; Nakeya. Utilization of Solar Energy in Liquid Milk Processing. *Magallat Buhut al-taqat al-Samsiyya.* **1985,** *3* (2), 35.

68. ReVelle, P., ReVelle, C. *The Global Environment: Securing a Sustainable Future;* Jones & Bartlett Learning: Burlington, 1992; p 480.

69. Reznik, D. Electro Heating Method. US Patent No 5863580, 1997; p 31.

70. Reznik, D. Ohmic Heating of Fluid Foods: Various Parameters Affect the Performance of Ohmic Heating Devices Used to Heat Fluid Food Products. *Food Technol.* **1996,** *5,* 251–260.

71. Robert, A. M.; Bishop, E. S. *Elements of Electricity;* American Technical Society: Chicago, IL, 1917; p 54.

72. Samaranayake, C. P.; Sastry, S. K. Electrode and pH Effects on Electrochemical Reactions during Ohmic Heating. *J. Electroanal. Chem.* **2005,** *577,* 125–135.

73. Sandu, C.; Lund, D. Fouling of Heat Exchangers: Optimum Design and Operation. In *Fouling of Heat Exchanger Surfaces;* Engineering Foundation: New York, 1983; pp 681–716.

74. Sastry, S. K.; Palaniappan, S. Mathematical Modeling and Experimental Studies on Ohmic Heating of Liquid-particle Mixture in a Static Heater. *J. Food Process Eng.* **1992,** *15,* 241–261.

75. Sastry, S. K. Ohmic Heating. *Encyclopedia Agric. Food Biol. Eng.* **2007,** *1* (1), 707–711.

76. Shimada, K.; Shimahara, K. Factors Affecting the Surviving Fractions of Resting *Escherichia coli* B and K-12 Cells Exposed to Alternating Current. *Agric. Biol. Chem.* **1981,** *45,* 1589–1595.

77. Shimada, K.; Shimahara, K. Responsibility of Hydrogen Peroxide for the Lethality of Resting *Escherichia coli* B Cells Anaerobically Exposed to an Alternating Current in Phosphate Buffer Solution. *Agric. Biol. Chem.* **1982,** *46,* 1329–1337.

78. Shimada, K.; Shimahara, K. Sublethal Injury to Resting *Escherichia coli* B Cells Aerobically Exposed to Alternating Current. *Agric. Biol. Chem.* **1983,** *47,* 129–131.

79. Shimada, K.; Shimahara, K. Changes in Surface Charge, Respiratory Rate and Stainability with Crystal Violet of Resting *Escherichia coli* B Cells Anaerobically Exposed to an Alternating Current. *Agric. Biol. Chem.* **1985,** *49,* 405–411.

80. Shimada, K.; Shimahara, K. Leakage of Cellular Contents and Morphological Changes in Resting *Escherichia coli* B Cells Exposed to an Alternating Current. *Agric. Biol. Chem.* **1985,** *49,* 605–607.

81. Shirsat, N.; Lyng, J. G.; Brunton, N. P.; McKenna, B. Ohmic Processing: Electrical Conductivities of Pork Cuts. *Meat Sci.* **2004,** *67,* 507–514.

82. Simmons, M., Jayaraman, P.; Fryer, P. The Effect of Temperature and Shear Rate upon the Aggregation of Whey Protein and its Implications for Milk Fouling. *J. Food Eng.* **2007,** *79,* 517–528.

83. Singh, R. P.; Hedman, D. R. A Volume of Food Science and Technology. *Introduction to Food Engineering*, 4th ed.; Academic Press Publication: Cambridge, MA, 2009; p 412.

84. Skudder, P. J. Ohmic Heating: New Alternative for Aseptic Processing of Viscous Foods. *Food Eng.* **1988,** *60,* 99–101.

85. Standards Australia. *Food Microbiology method.* AS/NZS 1766, 2001; p 312.

86. Steele, J. History, Trends, and Extent of Pasteurization. *J. Am. Vet. Med. Assoc.* **2000,** *217* (2), 175–178.

87. Stirling, R. Ohmic Heating: New Process for the Food Industry. *Power Eng. J.* **1987,** *1,* 365–371.

88. Tamime, A. Y. *Milk Processing and Quality Management;* Blackwell Publishing Ltd.: Oxford, England, 2009; p 312, 978-1-405-145-30.

89. Teknotext, A. B. *Dairy Processing Handbook: Tetrapak Processing Systems AB;* Lund Publisher: Sweden, 1995; pp 263–278.

90. Valentas, K. J.; Rotestein, E.; Singh, R. P., Eds.; *Handbook of Food Engineering Practice*; CRC Press: Boca Raton, FL, 1997; p 638.

91. Vicente, A. A. In *Novel Technologies for the Thermal Processing of Foods.* Proceedings of Encontro de Quimica dos Alimentos: Alimentos Tradicionais, Alimentos Saudaveis Rastreabilidade (CD-ROM), Be Instituto Politecnico de Beja, 1997; pp 499–506.

92. Vikram, V. B.; Ramesh, M. N.; Prapulla, S. G. Thermal Degradation Kinetics of Nutrients in Orange Juice Heated by Electromagnetic and Conventional Methods. *J. Food Eng.* **2005,** *69,* 31–40.

93. Villamiel, M.; Fadso, R. L.; Cono, N.; Olano, A. Denaturation of β-lactoglobulin and Native Enzymes in the Plate Exchanger and Holding Tube Section During Continuous Flow Pasteurization of Milk. *Food Chem.* **1997,** *58* (1–2), 49–52.

94. Wald, M. Growing Use of Electricity Raises Questions on Supply; New York Times: Manhattan, 1990. http://www.nytimes.com/1990/03/21/business/growing-use-of-electricity-raises-questions-on-supply.html (accessed on May 28, 2016).

95. Walstren, P.; Guerts, T. I.; Noomen, A.; Jellem, A.; Van, B. *Dairy Technology: Principles of Milk Properties and Processes*; Marcel Dekker: New York, 1999; p 312.

96. Wang, W. C.; Sastry, S. K. Salt Diffusion into Vegetable Tissue as a Pre-treatment for Ohmic Heating: Determination of Parameters and Mathematical Model Verification. *J. Food Eng.* **1993,** *20,* 311–323.

97. Zareifard, M. R.; Ramaswamy, H. S.; Trigui, M.; Marcotte, M. Ohmic Heating Behavior and Electrical Conductivity of Two-phase Food Systems. *Innov. Food Sci. Emerg. Technol.* **2003,** *4,* 45–55.

**CHAPTER 2**

# ADVANCES IN MICROWAVE-ASSISTED PROCESSING OF MILK

S. S. SHIRKOLE and P. P. SUTAR*

*Department of Food Process Engineering, National Institute of Technology, Rourkela 769008, Odisha, India*

*\*Corresponding author. E-mail: sutarp@nitrkl.ac.in; paragsutar@gmail.com*

## CONTENTS

## ABSTRACT

The introductory part of this chapter includes importance of thermal processing of milk, existing heating methods used and their effects on different quality characteristics. Authors explored alternative methods for thermal processing of milk and milk products, and briefly discussed microwave-assisted processing with recent research findings. Further, they discussed the influence of microwave heating on physicochemical properties, microbial destruction, enzyme inactivation, and volatile components of milk. Finally, they concluded that microwave heating has potential to meet industrial needs as an alternative technology for thermal processing of milk.

## 2.1   INTRODUCTION

Milk contains carbohydrates, fats, proteins, and vital minerals; and is consumed by all segments of the population. An escalation in production of milk, continuous demand for milk and milk products, and its perishable nature are some of the reasons that make processing of milk essential. Consumer demands for various types of processed milk and milk products. Industrial requisite for rapid production, enhanced product quality and shelf-life extension have brought notable developments in milk processing. Remarkable improvements have been focused toward milk separation, pasteurization, standardization, and packaging.[19]

   Shelf-life extension without negotiating product quality and safety has been a key objective of milk processors. Generally, application of heat is a common practice in milk processing industry in order to ensure microbial stability. However, noticeable changes in thermal processing have been observed. The technological developments together with the exertions and attentiveness of processors in improving quality of products have been motivating the researchers to develop innovative technology for milk processing. During milk processing operations, the alterations in different quality characteristics are the base to generate strong relationship between the quality of processed product and operating parameters. This is particularly useful for thermal treatment, where many heat-induced changes in milk components decide the functional properties of the processed product.

   Thermal technologies like dielectric heating, ohmic heating, and inductive heating have been developed as alternatives for the traditional methods of heating by conduction, convection, and radiation. Conventional thermal processing, which includes sluggish heat transfer from the source of heating to the cold spot, often results in heating of the product at the container periphery that is far more severe than that necessary to accomplish commercial sterility.[27]

Microwave-assisted thermal processing (MWATP) is an appropriate way as it is assumed to be fast, it results in to volumetric heating of product, and is clean, and easy to use technology. For thermal unit operations, application of microwaves is the subject of research since several years ago. Hamid et al.[21] were the first group to use the microwave heating for milk pasteurization. However, thermal processing by microwaves depends on the dielectric properties of the material to be heated.[32,47] Research on MWATP of milk is mainly focused on microbial destruction,[4] enzyme inactivation,[26] physicochemical properties,[23,46] and sensory changes.[45]

The purpose of this chapter is to focus on recent advances in MWATP of milk and its effects of microwaves on microbiological, physicochemical, and organoleptic properties of processed milk and milk products.

## 2.2 THERMAL PROCESSING OF MILK

The thermal processing and milk pasteurization are based on principles from research outcomes of Louis Pasteur (1822–1895). He developed a technique to prevent abnormal fermentation in wine by microbial destruction with application of heat to 60°C. After 100 years, the International Dairy Federation (IDF) defined pasteurization as, *"Pasteurization is a process applied to a product with the objective of minimizing possible health hazards arising from pathogenic microorganisms associated with milk by heat treatment which is consistent with minimal chemical, physical and organoleptic changes in the product."* The time–temperature combination of thermal treatment decides the final quality of product. Quality alteration can take place during and after thermal processing of milk, which changes its organoleptic, nutritional, physicochemical properties, and can also result in interactions between its principal constituents.[17] Additional effect due to technological treatment of milk is the increase in concentration of free fatty acids (FFAs).[6,28] While in thermal processing of milk, considerable alterations in physical properties of milk lipids can occur, especially at the level of fat globule membrane which is intricate in structure and easily ruptured by physical or thermal shock.[30] It results in excessive accumulation of FFA, normally associated with undesirable flavors.

In the most popular pasteurization process, heat is typically applied at the temperature (in the range of 60–80°C) below the boiling point of water for predetermined time interval. Also, other pasteurization methods such as ultra high temperature (UHT) pasteurization and high temperature short time (HTST) pasteurization are extensively used in the dairy industry. UHT sterilization process heats milk in the range 138–150°C for 4–15 s. Commercial sterilization of milk is successfully achieved by direct application of steam, such as steam infusion, or steam injection in which milk is quickly heated to 140°C by direct mixing of steam, followed by instantaneous cooling.[19]

HTST pasteurization is also used for extension of shelf-life of milk[29,36] and to improve keeping quality of milk without alterations in original characteristics by destructing the harmful microorganisms to acceptable level.[22] HTST pasteurization is capable of extending the shelf-life of milk up to 3 weeks, when milk is stored at refrigerated storage conditions and also depends upon the initial microbial load and level of refrigeration.[40] HTST pasteurization is generally carried out by the heating systems, which transfers the heat to the product by conduction and convection such as plate heat exchanger and tubular heat exchangers. The plate heat exchangers are widely used for heating as well as cooling in milk processing operations as they provide high degrees of effectiveness and compactness.[18] However, fouling and deposit of minerals, and proteins on heat exchanger surfaces are crucial issues in both UHT and HTST pasteurization systems.[24] Fouling affects economy of milk processing plant as it reduces efficiency of heat transfer, increase in pressure drop. Possibility of product deterioration increases as a result of fouling, because the required pasteurization temperature cannot be reached by the product during processing.[8]

Thermal technologies mentioned above are well established in milk processing industry while new heating methods by microwave, ohmic, and radiofrequency are under development.[4,7] They all have common heat generation feature and show robust effects in terms of heat and energy efficiency. Microwave-assisted pasteurization of milk is under focus as it is quick, clean, and easy to handle also offers volumetric heating, high thermal efficiency, and can be applicable to continuous pasteurization systems.

## 2.3    MICROWAVE HEATING OF MILK

To minimize drawbacks associated with conventional thermal processing,[27] microwave heating is a potential alternative. Microwaves are electromagnetic waves within 300 MHz to 300 GHz frequency band. Microwave energy penetrates into milk and generates volumetric heat due to dipole action of solvent and conductive movement of dissolved ions. The dipole action is caused due to variations in magnetic and electric field. Also, water is the key source for interaction with microwave because of its dipole nature.

*Specific features of microwave heating*

- Microwave heating is a volumetric heating phenomenon throughout the entire volume of product, uniform and quick, effectively reduces process time, and energy consumption.
- Nutrients and other sensory characteristics of microwave-processed milk are well preserved due to rapid heat generation and short process time.
- Minimum fouling depositions due to elimination of hot contact surfaces, since piping used for microwave pasteurization system are transparent to waves and remains comparatively cooler than inside hot milk.
- Perfectly suitable for clean-in-place (CIP) system.
- Microwave heating provides very easy control during operation. It can be switched on/off quickly.
- Pasteurization of the milk by microwave is possible even after bottling and packaging.

These features can yield enhanced product quality and increased productivity.[35,43] The studies on microwave-assisted pasteurization and sterilization of milk as well as physicochemical, thermal, and sensory properties of milk have been conducted by researchers.[12,14,20,37]

## 2.3.1    FACTORS AFFECTING MICROWAVE HEATING OF MILK

### 2.3.1.1    MOISTURE CONTENT AND TEMPERATURE

Free water content in the product determines the heating of product in microwave oven.[31,33] Higher percentage of water content transforms electromagnetic energy into greater absorption of microwave. When moisture content is high the product heats efficiently as it has high dielectric loss factor.[38] Dielectric properties are dependent on chemical composition and dipole moment associated with water present in product. The dielectric properties of milk and its constituents at 20°C and 2.44 GHz microwave frequency were studied by Kudra et al.[25] The dielectric constant may decrease or increase with temperature of the product.

### 2.3.1.2   COMPOSITION

The dielectric properties are determined by composition of the milk as well as significantly influenced by contents of water and salt especially at 450 and 900 MHz, respectively.[34] The dielectric properties for fresh cow milk with 70–100% concentrations stored at 22°C for 36 h and 5°C for 144 h (measured from 10 to 4500 MHz frequencies) have been reported by Guo et al.[20] They concluded that loss factor was minimum at 1700 MHz and dielectric constant of milk decreased with increase in frequency. Also, they found that penetration depth increased with water content and storage time. The dielectric properties have been listed for different processed cheeses at 20 and 70°C temperatures studied by Datta and Nelson.[15] They observed that loss factor increases slightly with temperature at higher moisture and lower fat content.[15] The aqueous nonfat dry milk solutions were used as model systems to measure dielectric properties by To et al.[41] They reported that milk salts and ash content affect the dielectric properties.

### 2.3.1.3   MICROWAVE FREQUENCY

The penetration depth of microwave is greatly influenced by its frequency.[14] For microwave heating, only two frequencies (2450 and 915 MHz) are allocated. Then, 2450 MHz is approved for commercial microwave ovens and 915 MHz is used for industrial purpose. Microwaves at 2450 and 915 MHz can travel at 0.12 and 0.33 wavelengths, respectively.[5] The relationship between penetration depth and wavelength is expressed below:

$$D = \frac{\lambda_0 (\varepsilon')^{1/2}}{2\pi\varepsilon''} \qquad (2.1)$$

where $D$ is penetration depth (at which 63% of incident energy is absorbed), $\lambda_0$ is wavelength in free space, $\varepsilon'$ is relative dielectric constant, and $\varepsilon''$ is relative dielectric loss factor.

As wavelength ($\lambda$) increases (frequency decreases), penetration depth also increases. The dielectric loss factor is highly affected by frequency.

### 2.3.1.4   PRODUCT PARAMETERS: MASS, DENSITY, AND GEOMETRY

There is linear relationship between dielectric constant and density of the product influencing the microwave heating. In low-density products the air present in pores makes it good insulator.[39] The size and shape of the product has great influence on heating pattern during microwave heating. Generally, products with corners (90° edges) have tendency of localized heating due to concentration of microwave energy from multi directions.[9] Microwave heating generates highest temperature and hot spot in the centers of the sphere. While in cylindrical objects, microwave leads to non-uniform distribution of energy with elevated temperatures at center. In brick and cube shaped objects, microwave energy concentrates in the corners, leads to hot spots resulting in to nonuniform heating.

## 2.4   EFFECTS OF MICROWAVE HEATING ON MILK

Milk may be considered as an emulsion of fat in an aqueous solution which are both colloidal in nature. Milk contains inorganic components such as P, Ca, Na, Mg, Cl, K, etc., and organic components such as fat, vitamins, protein, lactose, and enzymes. During MWATP, milk can undergo considerable changes in physicochemical and microbiological properties.[13]

### 2.4.1   EFFECT ON PHYSICOCHEMICAL PROPERTIES

Influence of microwave heating on the physicochemical characteristics of cow milk was studied by Iuliana et al.[23] In order to preserve milk quality, different physicochemical parameters were considered by Iuliana et al.[23] with respect to microwave exposure time is given. They found that values of dry matter in microwave-exposed samples were decreased. It was due to decrease in protein, fat, and lactose as well as some vitamins such as $B_9$, A, E, and C. Similar results were reported by Dumuta et al.[16] Fat content in microwave treated milk decreased from 4.73 to 3.00% after 120 s heating time indicated 36.58% decrease in fat. Specific gravity of milk varied from 1.031 to 1.033 for 120 s exposure of milk to microwaves. The increase in specific gravity indicates that there has been evaporation of moisture from milk during microwave heating.

The influence of milk protein content on the surface tension and viscosity was studied by Caprita et al.[11] They found that the protein content in microwave treated milk samples were decreased from 3.48 to 3.34 g%. The sample with high protein content (> 3 g%) showed a viscosity of about 2 cP. The viscosity of milk was also considered as a product control parameter rather only quality factor by Al-Belaty.[3] Electrical conductivity is the property to transport electrical charges. In heterogeneous system such as milk, the colloidal dispersed elements and fat obstruct the ions during movement, which results in the decreased conductivity. The correlations between electrical conductivity of milk and lactose concentrations as well as surface tension and protein contents have been investigated during microwave heating of milk by several investigators.[10,23]

Effect of microwave treatment on composition of milk was investigated by Constantin and Csatlos.[13] They reported no significant alterations in the chemical structure when heated with microwave energy at high temperature. The changes were visible only at second decimal for lactose and protein composition, and at the third decimal for milk freezing point. Comparison between microwave and conventional pasteurization system and their effect on chemical and sensorial properties were investigated by Valero et al.[45] They treated raw milk using microwave-assisted and conventional heat exchangers under similar conditions at 80 or 92°C for 15 s and stored at 4.5 ± 0.5°C for 15 days. No qualitative changes were observed by them between conventional and microwave heated samples.

## 2.4.2   EFFECT ON VOLATILE COMPONENTS

Alterations in volatile components in conventionally and microwave-assisted heated milk were studied by Valero et al.[44] They selected 12 compounds and compared milk processed in both microwave and conventional heating systems at 70–90°C. Changes in volatile contents of milk, processed under both microwave and conventional methods have been studied by Valero et al.[44] Volatile components were identified using gas chromatography and mass spectroscopy (GC-MS) and quantified by GC techniques. They concluded that composition of volatile content was same after both microwave and conventional treatment provided that there was satisfactory temperature control and uniform heating. Further they mentioned that microwave-assisted heating had no adverse effects on flavors. Volatile compounds like ketones, aldehydes, alcohols, esters, and aromatic hydrocarbons remains unchanged in freshly microwave pasteurized milk as that of found in raw milk.[45]

## 2.4.3 EFFECT ON MICROBIAL DESTRUCTION

There are few research findings on microwave-assisted milk pasteurization. Influence of microwave and indirect UHT treatment on microbiological quality of skim milk was compared by Clare et al.[12] Both microwave and UHT treatment inactivated all microbial loads in raw milk as proven by absence of colony forming units (CFU) using different microbial testes and sterility were maintained over a storage period of one year. They concluded that off flavor developed in UHT treated samples were typically related with the final products which affect the consumer acceptance. Therefore, MWATP can be alternative to UHT.

Microwave-assisted milk flash pasteurization and its effect on microbiological, chemical, and thermo-physical characteristics of milk were studied by Al-Hilphy and Ali.[4] They estimated total number of microbes and *Escherichia coli* and found that microbial quality of raw milk was improved after microwave-assisted flash pasteurization at 100°C for 0.01 s. It reduced total microbial count and ensured nonexistence of coliforms in pasteurized milk. Also, they reported significant reduction in thiobarbituric acid (TBA) and FFAs after flash pasteurization. Tremonte et al.[42] studied influence of microwave treatment, boiling, and refrigeration on microbial quality of milk. The microwave treatment at 900 W for 75 s results in destruction of microorganisms similar to that of milk treatment by boiling and maintaining the whey protein contents of milk. They showed that domestic boiling affects the nutritional quality of milk drastically. Finally, they concluded that microwave treatment could be alternative for domestic practice of milk boiling.

## 2.4.4   EFFECT ON ENZYME INACTIVATION

Effect of conventional heating and microwave-assisted continuous flow heating systems on inactivation of phosphatase was studied by Lin and Ramaswamy.[26] They exposed raw milk to microwave-assisted continuous flow heating, conventional isothermal water bath heating and continuous flow thermal system in the temperature range 60–75°C for pasteurization. The D-values of associated alkaline phosphatase (ALP) varied from 1.7 s at 75°C to 1250 s at 60°C with a z-value of 5.2°C under conventional batch heating, 13.5 s at 70°C to 128 s at 65°C with a z-value of 5.2°C under continuous flow thermal holding, and 1.7 s at 70°C to 17.6 s at 65°C with a z-value of 4.9°C under continuous flow microwave heating. Their results revealed that ALP inactivation occurs very faster under microwave-assisted heating than conventional heating and confirming microwave pasteurization can diminish severity of treatment and offers higher quality pasteurized milk. Microwave-assisted milk flash pasteurization at temperature of 100°C for 0.01 s showed absence of alkaline phosphatase after pasteurization treatment as investigated by Al-Hilphy and Ali.[4]

## 2.5   OTHER APPLICATIONS OF MICROWAVE PROCESSING IN MILK

## 2.5.1   MICROWAVE-ASSISTED THAWING

Effect of microwave-assisted thawing and gravitational methods were investigated on cryoconcentration to produce milk whey concentrates by Aider et al.[2] Their results indicated that cryoconcentration is an effective method to produce milk whey protein. Also, they observed that microwave-assisted thawing was effective as compared to gravitational thawing.

## 2.5.2   MICROWAVE SENSOR

The usage of a moisture sensor with microwave absorption at 40 GHz was explored to determine the water content in milk by Agranovich et al.[1] Since microwave sensor directly analyzes free water content in the medium, it can replace near infrared (NIR) spectrometer and pH meter in monitoring the process with high precision. Also, they reported that microwave measurements at various frequency levels would increase the accuracy and may be helpful to get additional facts about milk quality.

## 2.6  CONCLUSIONS

Application of heat for thermal processing of milk is a comparatively robust preservation technology but still numerous challenges exist for development to meet the industrial needs. Microwave processing has several advantages as compared to the conventional methods used for thermal processing of milk. It can afford new challenges for shelf-life extension with better quality retention, microbial stable product, and energy efficiency. Due to continuous developments in microwave equipment manufacturing practices and process, the milk processing using microwaves can be a promising and alternative technology for traditional thermal processing of milk.

## KEYWORDS

- clean-in-place
- colony forming units
- cryoconcentration
- enzyme inactivation
- fermentation
- HTST pasteurization
- spray drying
- thawing
- thermal processing
- UHT pasteurization

## REFERENCES

1. Agranovich, D.; Renhart, I.; Ishai, P. B.; Katz, G.; Bezman, D.; Feldman, Y. A. Microwave Sensor for the Characterization of Bovine Milk. *Food Control.* **2016,** *63,* 195–200.

2. Aider, M.; De Halleux, D.; Melnikova, I. Gravitational and Microwave-Assisted Thawing during Milk Whey Cryoconcentration. *J. Food Eng.* **2008,** *88* (3), 373–380.
3. Al-Belaty, S. *Quality Control and Food Standard Specifications;* Press and Publishing Dar Alhekma: Ministry of Higher Education and Scientific Research, Musol University, Iraq, 1988; p 112.
4. Al-Hilphy, A. R. S.; Ali, H. I. Milk Flash Pasteurization by the Microwave and Study Its Chemical, Microbiological and Thermo Physical Characteristics. *J. Food Process. Technol.* **2013,** *4,* 250.
5. Annis, P. J. Design and Use of Domestic Microwave Ovens. *J. Food Prot.* **1980,** *43* (8), 629–632.
6. Antonelli, M.; Curini, R.; Scricciolo, D.; Vinci, G. Determination of Free Fatty Acids and Lipase Activity in Milk: Quality and Storage Markers. *Talanta.* **2002,** *58* (3), 561–568.
7. Awuah, G.; Ramaswamy, H.; Economides, A.; Mallikarjunan, K. Inactivation of *Escherichia coli* K-12 and *Listeria innocua* in Milk Using Radio Frequency (RF) Heating. *Inn. Food Sci. Emerg. Technol.* **2005,** *6* (4), 396–402.
8. Bansal, B.; Chen, X. D. A Critical Review of Milk Fouling in Heat Exchangers. *Compr. Rev. Food Sci. Food Safety.* **2006,** *5* (2), 27–33.
9. Campanone, L.; Zaritzky, N. Mathematical Analysis of Microwave Heating Process. *J. Food Eng.* **2005,** *69* (3), 359–368.
10. Caprita, A.; Caprita, R. The Effect of Lactose Content on the Milk Electric Conductivity. *Ann. West Univ. Timisoara Series Chem.* **2001,** *10* (2), 375–378.
11. Caprita, R.; Caprita, A.; Benscik, I.; Cretescu, I. The Influence of Milk Protein Content on the Surface Tension and Viscosity. *Acta Vet. Scand. Suppl.* **2003,** *98,* 263.
12. Clare, D.; Bang, W.; Cartwright, G.; Drake, M.; Coronel, P.; Simunovic, J. Comparison of Sensory, Microbiological, and Biochemical Parameters of Microwave Versus Indirect UHT Fluid Skim Milk during Storage. *J. Dairy Sci.* **2005,** *88* (12), 4172–4182.
13. Constantin, A.; Csatlos, C. Research on the Influence of Microwave Treatment on Milk Composition. *Bull. Transilvania Univ. Braşov.* **2010,** *3,* 52.
14. Coronel, P.; Simunovic, J.; Sandeep, K. Temperature Profiles within Milk after Heating in a Continuous-Flow Tubular Microwave System Operating at 915 MHz. *J. Food Sci.* **2003,** *68* (6), 1976–1981.
15. Datta, A. Ed.; *Handbook of Microwave Technology for Food Applications*; CRC Press: Boca Raton, FL, 2001; p 536.
16. Dumuta, A.; Giurgiulescu, L.; Mihaly-Cozmuta, L.; Vosgan, Z. Physical and Chemical Charateristics of Milk. Variation Due to Microwave Radiation. *Croat. Chem. Acta.* **2011,** *84* (3), 429–433.
17. Fox, P. F.; McSweeney, P. L. *Dairy Chemistry and Biochemistry;* Blackie Academic and Professional: London, 1998; p 231.
18. Ghosh, I.; Sarangi, S.; Das, P. An Alternate Algorithm for the Analysis of Multistream Plate fin Heat Exchangers. *Int. J. Heat Mass Tran.* **2006,** *49* (17), 2889–2902.
19. Goff, H.; Griffiths, M. Major Advances in Fresh Milk and Milk Products: Fluid Milk Products and Frozen Desserts. *J. Dairy Sci.* **2006,** *89* (4), 1163–1173.
20. Guo, W.; Zhu, X.; Liu, H.; Yue, R.; Wang, S. Effects of Milk Concentration and Freshness on Microwave Dielectric Properties. *J. Food Eng.* **2010,** *99* (3), 344–350.

21. Hamid, M.; Boulanger, R.; Tong, S.; Gallop, R.; Pereira, R. Microwave Pasteurization of Raw Milk. *J. Microw. Power.* **1969**, *4* (4), 272–275.
22. Hudson, A.; Wong, T.; Lake, R. *Pasteurisation of Dairy Products: Times, Temperatures and Evidence for Control of Pathogens;* Institute of Environmental Science and Research Ltd: Christchurch, New Zealand, 2003; p 62.
23. Iuliana, C.; Rodica, C.; Sorina, R.; Oana, M. Impact of Microwaves on the Physico-Chemical Characteristics of Cow Milk. *Rom. Rep. Phys.* **2015**, *67* (2), 423–430.
24. Johansson, C. *Heating of Milk and Milk Fouling.* Technical Report: Linköpings Universitet – Projekt i mikrobiell bioteknik TVMB04: www.liu.se/content/1/c6/11/88/57/2008/Cecilia_Johansson.pdf. May 15, 2008.
25. Kudra, T.; Raghavan, V.; Akyel, C.; Bosisio, R.; Van de Voort, F. Electromagnetic Properties of Milk and Its Constituents at 2.45 GHz. *J. Microw. Power Electromagn. Energy.* **1992**, *27* (4), 199–204.
26. Lin, M.; Ramaswamy, H. S. Evaluation of Phosphatase Inactivation Kinetics in Milk under Continuous Flow Microwave and Conventional Heating Conditions. *Int. J. Food Prop.* **2011**, *14* (1), 110–123.
27. Meredith, R. J. *Engineers' Handbook of Industrial Microwave Heating;* The Institution of Engineering and Technology (IET), Stevenage Herts, UK; 1998; Vol. 25, pp 19–50.
28. Morgan, F.; Gaborit, P. The Typical Flavor of Goat Milk Products: Technological Aspects. *Int. J. Dairy Technol.* **2001**, *54* (1), 38–40.
29. Morr, C. Effect of HTST Pasteurization of Milk, Cheese Whey and Cheese Whey UF Retentate upon the Composition, Physicochemical and Functional Properties of Whey Protein Concentrates. *J. Food Sci.* **1987**, *52* (2), 312–317.
30. Muir, D. Milk Chemistry and Nutritive Value. In *Dairy Products;* R. Early, ed.; Blackie Academic & Professinal: London, 1988; p 354.
31. Nelson, S.; Kraszewski, A. Grain Moisture Content Determination by Microwave Measurements. *Trans. ASAE.* **1990**, *33* (4), 1303–1305.
32. Nunes, A.; Bohigas, X.; Tejada, J. Dielectric Study of Milk for Frequencies between 1 and 20 GHz. *J. Food Eng.* **2006**, *76* (2), 250–255.
33. Ohlsson, T. Fundamentals of Microwave Cooking. *Microw. World.* **1983**, *4* (2), 4–9.
34. Ohlsson, T.; Bengtsson, N. Dielectric Food Data for Microwave Sterilization Processing. *J. Microw. Power.* **1975**, *10* (1), 93–108.
35. Pereira, R. N.; Vicente, A. A. Novel Technologies for Milk Processing. In *Dairy Products;* R. Early, ed.; Blackie Academic & Professinal: London, 1988; Chapter 7, p 354.
36. Raynal-Ljutovac, K.; Park, Y.; Gaucheron, F.; Bouhallab, S. Heat Stability and Enzymatic Modifications of Goat and Sheep Milk. *Small Rumin. Res.* **2007**, *68* (1), 207–220.
37. Salazar-González, C.; San Martín-González, M. F.; López-Malo, A.; Sosa-Morales, M. E. Recent Studies Related to Microwave Processing of Fluid Foods. *Food Bioprocess Technol.* **2012**, *5* (1), 31–46.
38. Schiffmann, R. Food Product Development for Microwave Processing. *Food Technol. USA.* **1986**, *40*, 94–98.
39. Schiffmann, R. Microwave Foods: Basic Design Considerations. *Tappi J.* **1990**, *73* (3), 209–212.

40. Sepulveda, D.; Góngora-Nieto, M.; Guerrero, J.; Barbosa-Cánovas, G. Production of Extended Shelf-life Milk by Processing Pasteurized Milk with Pulsed Electric Fields. *J. Food Eng.* **2005,** *67* (1), 81–86.

41. To, E. C.; Mudgett, R. E.; Wang, D. I.; Goldblith, S.; Decareau, R.; Dielectric Properties of Food Materials. *J. Microw. Power.* **1974,** *9* (4), 303–315.

42. Tremonte, P.; Tipaldi, L.; Succi, M.; Pannella, G.; Falasca, L.; Capilongo, V.; Coppola, R.; Sorrentino, E. Raw Milk from Vending Machines: Effects of Boiling, Microwave Treatment, and Refrigeration on Microbiological Quality. *J. Dairy Sci.* **2014,** *97* (6), 3314–3320.

43. USFDA. Kinetics of Microbial Inactivation for Alternative Food Processing Technologies: Microwave and Radio Frequency Processing. http://www.fda.gov/Food/FoodScienceResearch/SafePracticesforFoodProcesses/ucm 100250.htm (accessed June 5, 2015).

44. Valero, E.; Sanz, J.; Martinez-Castro, I. Volatile Components in Microwave and Conventionally Heated Milk. *Food Chem.* **1999,** *66* (3), 333–338.

45. Valero, E.; Villamiel, M.; Sanz, J.; Martınez-Castro, I. Chemical and Sensorial Changes in Milk Pasteurized by Microwave and Conventional Systems during Cold Storage. *Food Chem.* **2000,** *70* (1), 77–81.

46. Villamiel, M.; López-Fandiño, R.; Corzo, N.; Martínez-Castro, I.; Olano, A. Effects of Continuous Flow Microwave Treatment on Chemical and Microbiological Characteristics of Milk. *Z. Lebensm. Unters. Forsch.* **1996,** *202* (1), 15–18.

47. Wang, W.; Chen, G.; Gao, F. Effect of Dielectric Material on Microwave Freeze Drying of Skim Milk. *Dry. Technol.* **2005,** *23* (1–2), 317–340

# CHAPTER 3

# ADVANCEMENT IN THE GHEE MAKING PROCESS

PRAMOD KUMAR[1], RAMAN SETH[1], HITESH PATEL[1], ANIT KUMAR[2], DURGA SHANKAR BUNKAR[3], and MOHAMMED NAYEEM[2*]

*[1]Dairy Chemistry Division, National Dairy Research Institute (NDRI), Karnal 132001, Haryana, India*

*[2]Department of Food Science and Technology, National Institute of Food Technology Entrepreneurship and Management, Sonipat 131028, Haryana, India*

*[3]Centre of Food Science and Technology, Institute of Agricultural Sciences, Banaras Hindu University, Varanasi 221005, India*

*[*]Corresponding author. E-mail: nayeem.amaan@gmail.com*

## CONTENTS

## ABSTRACT

In the latest scenario, modern technology-assisted plants have the capacity of 400–500 kg/h of ghee, using creamy butter or high fat cream as raw material. This system offers a number of advantages, namely, compactness in design, hygienic operation, clean-in-place (CIP) cleaning, reduced strain over the operator, absence of fouling and foaming, short residence time, and hence, the final product contains higher percentage of vitamins as compared to ghee made in kettles. Apart from these benefits, the continuous ghee making system incorporates an excellent feature of energy used for preheating butter/cream. An average saving of 0.17 kg steam/kg ghee is achieved. If this system of ghee manufacturing is adopted in the organized dairy sector, an annual saving of furnace oil works out to be approximately 4 million liters.

## 3.1  INTRODUCTION

Ghee is one of the most common and ancient dairy product in India. *"Ghee is a class of clarified butter that originated in ancient India and is commonly used in South Asian, Iranian, and Arabic cuisines, traditional medicine, and religious rituals,"* according to https://en.wikipedia.org/wiki/Ghee. It is prepared from either cow or buffalo milk. Ghee is a highly demanded milk product throughout the world, especially in India. It is prepared by heating cream up to a particular stage to remove moisture that inhibit the growth of micro-organisms and increase the shelf-life. It has an economical value to the dairy industry that can be proved from the fact that more than 30% of milk produced in India is converted to ghee. This is because preparation of ghee is not very complicated and technical so that people can easily prepare it at their homes.

Chemically, ghee is a complex mixture of many fatty acids and glycosides, phospholipids, fat-soluble vitamins, and carotenoids (cow ghee), therefore, it does not have a sharp melting point. Ghee also contains traces of iron, calcium, and moisture. The major constituent of ghee is glycerides, which constitute 98% of total material in ghee and the rest consists of 0.5% cholesterol. The physical, chemical, and nutritional qualities of ghee depend upon the quality of raw materials used, like milk, cream, and butter. Storage condition of the ghee will decide its taste, flavor, and shelf-life and further quality of raw material depends on the type of feed given to the cattle. The quality of ghee is determined by various physical testing parameters like: peroxide value, flavor, acidity, and Rancimat test.

Generally, preparation of ghee is made at a temperature range of 105–110°C, and above this temperature, the ghee gives a very intense and burnt flavor which is not liked by the consumers and also it have lower shelf-life due to chances of oxidation. When the ripened cream has acidity in the range of 0.15–0.25%, its ghee has a good flavor and taste. Rate of rancidity development in the stored ghee is less prepared from unripened butter as compared to ghee made from ripened butter. Formation of bitter taste of small molecules of aldehyde, ketone, free fatty acids and simple glycerides due to action of moisture, oxygen, and lipase enzyme is called hydrolytic, oxidative, and lipolytic rancidity, respectively. Other enzymes involved in the oxidative and enzymatic rancidity are lipoxidase enzymes, dehydrogenase enzymes, and decarboxylase enzymes.

Ghee price is increasing day by day as it has become a special part of daily diet because of its good flavor and nutrition. There is no change in the preparation method of ghee from ancient to current times, regardless of the volume of production. It is manufactured as a batch process, which has many disadvantages and losses. The principles and operation of equipment for preparation of ghee at the cottage level are translated to industrial levels for its production in large quantity. Consequently, inefficient energy supply, unhygienic conditions, poor sanitation, and non-uniform product quality associated with the rural scale operation crept into the large-scale preparation of ghee in dairy plants. The serious problems associated with manufacturing of ghee in the industry are:

- Unsanitary operation, product exposed to environment, and increasing chances of contamination.
- Product spillage around the equipment, making the floor slippery and causing accidents.
- Low heat transfer coefficient causing bulky equipment.
- Formation of tenacious scale of ghee and milk residues on the heating surface, adding to poor performance of equipment and making cleaning and sanitation strenuous.
- Equipment and processes are unsuitable for large volumes of production.
- Large residence time and product inventory in the equipment presenting greater risk of bulk spoilage of the product.
- Excessive strain and fatigue on the operator.

Therefore, a demand for efficient, labor saving and sanitary processing of ghee exists in the dairy industry. Continuous processing of ghee overcomes several of these problems. Keeping in view the problems and limitations as stated above, new equipment was developed, which can make ghee continuously, instead of a batch process.

Unlike certain Western dairy products, Indian dairy products (such as cheese, sweetened condensed milk, rasogolla, or rabri) with which only a few amongst us have professional acquaintance, ghee, and its manufacture appears to be familiar to most of us and Indian dairymen, through our professional as well as personal experience. For this reason, it will not be necessary to detail here the most obvious aspects concerning the industrial ghee production trends and innovations.

## 3.2   ADVANCED METHOD OF GHEE MAKING

### 3.2.1   CONTINUOUS GHEE MAKING METHOD BASED ON FLASH EVAPORATION

The flow diagram of this process is shown in Figure 3.1. It consists of vat for receiving and heating of cream and gravity separator (it is optional for making ghee from cream) with high pressure scraped surface heat exchangers attached with vapor separator and positive displacement pumps (PAs) to forward the raw material to different units. The raw material (cream or butter) is received in a jacketed receiving vat (1) for melting. When butter is the raw material, PA will pump part of the melted butter to gravity separator (2) and part to the balance tank (3A). The gravity separator (2) separates the butter into two portions: one fat rich component and the other, buttermilk. The fat rich fluid containing a little buttermilk goes to the balance tank (3A) through the outlet (2LF) and buttermilk comes out of gravity separator through outlet 2LB. In case of cream, the gravity separator (2) is by-passed and PA sends the cream directly to the balance tank (3A).

The gravity separator (2) can also be by-passed when using butter, and in that case, the raw material is handled in the same manner as in case of cream. The pump (PB) receives the raw material from balance tank (3A) and sends it through the first scraped surface heat exchanger (4A) to vapor separator (5A). In the heat exchanger, the raw material is heated with the

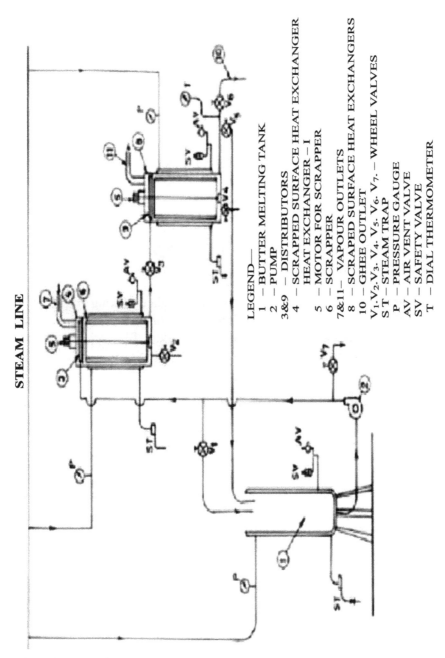

LEGEND—
1 – BUTTER MELTING TANK
2 – PUMP
3&9 – DISTRIBUTORS
4 – SCRAPPED SURFACE HEAT EXCHANGER
HEAT EXCHANGER – I
5 – MOTOR FOR SCRAPPER
6 – SCRAPPER
7&11– VAPOUR OUTLETS
8 – SCRAPED SURFACE HEAT EXCHANGERS
10 – GHEE OUTLET
$V_1$, $V_2$, $V_3$, $V_4$, $V_5$, $V_6$, $V_7$ – WHEEL VALVES
ST – STEAM TRAP
P – PRESSURE GAUGE
AV – AIR VENT VALVE
SV – SAFETY VALVE
T – DIAL THERMOMETER

**FIGURE 3.1** Continuous ghee making plant.

help of steam and then the super-heated liquid (cream/butter) is flashed in the vapor separator (5A). The vapor separator (5A) separates the water vapor from the liquid fat. The fluid from the separator (5A) with partially removed moisture goes to the balance tank (3B) where it is pumped to the heat exchanger (4B), and then to the vapor separator (5B). If butter is taken as raw material, the process of complete removal of water and flavor development is completed in the second stage. However, for cream, a third stage involving balance tank (3C), heat exchanger (4C), and vapor separator (5C) may be used to have a better control on the quality of the products. The sediment (ghee residue) in the product coming out of final stage may be removed by any of the standard filtering or clarifying devices, before sending the product for packaging.

### 3.2.2   COMPACT DESIGN OF CONTINUOUS GHEE MAKING FALLING FILM PRINCIPLE

Abichandani et al. indicated that continuous ghee making based on the principle of flash evaporation can be made compact and economically viable.[1] The new plant was designed on falling film principle with a capacity of 100 kg/h. The sequence of steps is as follows:

- Tap water is filled in the jacketed tank. The pump and scrapper motors are switched on. All the valves except V7, V2, and V4 are in open position. Water is circulated for 5–10 min.
- Detergent solution (washing soda + teepol) of 0.5% strength at temperature of 75–80°C for 10 min.
- The plant is rinsed with hot water at 75–80°C for 10 min.
- The heat exchangers are drained by opening valves V2, V4, and V7.
- The valves V7, V2, V4, and V6 are closed and valves V1 and V5 are opened.
- Scraper motors are switched off.
- Butter is put into the butter-melting tank. Steam pressure in the jacket is kept at 1.0 kg/cm².
- After melting butter, the valve V1 is adjusted to the desired flow rate and the scraper motors are switched on.
- The product is recirculated for 20–30 min. During this period, the plant gets warmed up.

- When desired temperature of ghee is attained (as indicated by dial thermometer), recirculating valve V5 is closed and outlet valve V6 is opened.
- In order to maintain thermal equilibrium, the level of butter in butter tank should be kept constant as far as possible.

At the end of operation, the plant is cleaned with hot water and detergent and then finally rinsed with hot water. A trial run with water was conducted to know the rate of evaporation. Water at 40°C was pumped into the heat exchanger for the removal of moisture at the rate of 160 kg/h. The steam pressure in the jacket was adjusted to 2.0 kg/cm². The rate of evaporation was found to be 80 kg/h. The economy of operation was 0.60 kg/kg of steam. It was reported that the economy of operation in conventional ghee pan operated under similar conditions is 0.35 kg/kg steam. Thermal efficiency of the plant was found to be 71% as compared to 37% obtained in ghee pan. Even if the heat energy equivalent to electrical energy is taken into account, the overall efficiency of the plant was worked out to be 44%.

### 3.2.3   CONTINUOUS GHEE WITH ENERGY CONSERVATION

Figure 3.2 illustrates the schematic hook-up of continuous ghee making based on energy conservation. Butter is heated to 70–75°C in tank (1) with steam. The pump (2) is started and the flow rate of molten butter is adjusted by valve V1 to the desired value, as indicated in rotameter (3).

The rotor drive is switched on to admit the steam into the jacket (4) of thin film scraped surface heat exchanger. The centrifugal action of the rotor (6) spread the product uniformly as a thin film on the heating surface (5). The rate of evaporation of moisture from product film is very rapid due to the thin film of product and the turbulence induced by the blade action.

The vapors are removed through vapor outlet (9) and vapors from the outlet can be used for preheating butter in the tank (1A) for steam economy. During the period of warm-up, the partially concentrated fat is diverted into the balance tank by keeping the valves V3 open and V2 closed. The temperature indicators T1 and T2 indicate the temperature of molten butter and ghee, respectively. On ghee attaining the desired temperature, valve V3 is closed and valve V2 is opened. Ghee is collected in tank (7). The

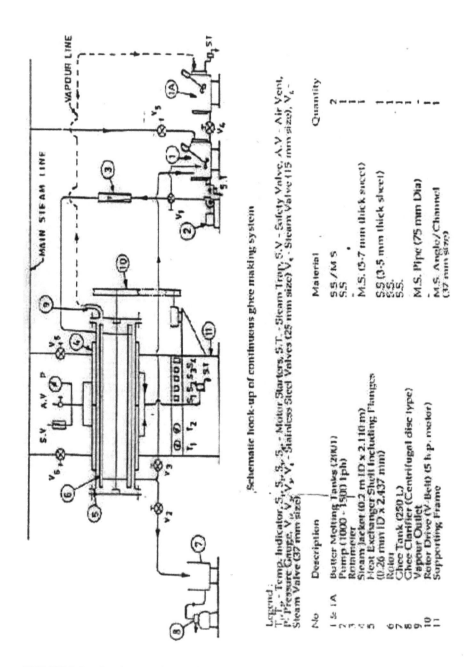

**FIGURE 3.2**  Continuous ghee making plant with energy conservation.

residue is separated by clarifier (8). The plant can also make ghee from high fat cream. Salient features of this design are:

- High heat transfer coefficients.
- No fouling and foaming problems.
- Short residence time in heated zone.
- High capacity reduction.
- Low product inventory and no chances of bulk spoilage.
- Energy conservation is possible by heat recovery.
- Automation of clean-in-place (CIP).
- Minimum strain for the operator.
- Hygienic operation and better product quality.

## 3.3   OTHER ALTERNATIVE METHOD OF GHEE MAKING

Ghee is most commonly prepared in India as the home industry, and is usually prepared by three methods: (1) creamy butter method, (2) direct cream method, and (3) pre-stratification method. The direct cream method is most commonly used in villages. Preparation of ghee is done by following these steps (Fig. 3.3):

- During the process of pasteurization, milk cream is separated by using cream separator with 60–75 % fat content.
- The cream is churned with the help of churner, which facilitates the removal of water as whey liquid which increases the fat percentage up to 80–85%. The whey is drained in whey tank, which is obtained as a byproduct.
- Prepared butter is transferred to cold storage till sufficient butter is available for ghee making[2,3] in ghee kettle.
- Before processing of butter in the steam-jacketed kettle, all the condensates of steam are removed.
- Slowly, the steam valve is opened in the ghee kettle to start steam supplying (1.5 bars); this will slowly melt the butter without burning it.
- Tightly close the drain valve, open the steam tap and let the condensate pass through it.
- Once enough butter is melted, start the agitator to agitate melted butter for rapid removal of moisture.

- With continuous supply of steam, the butter oil will start boiling and water from the butter will evaporate.
- Let all the water evaporate.
- Slowly close the supply of steam valve and open the vent valve on the jacket.
- Let all condensate from the steam jacket get drained, and then cool down the ghee.
- Continue the agitator motion for agitating ghee.
- When ghee temperature is decreased to around 70°C, pass this to the ghee filter tank.
- Most of the burned protein from the butter is filtered out with the use of stainless steel strainer.
- Once all ghee is filtered, start the ghee pump.
- Ghee is clarified with the use of ghee clarifier and collected in the balance tank.
- Ghee from the balance tank is transferred to jacketed ghee storage tank with the help of another ghee pump.
- To keep ghee in free flowing condition, this storage tank has water jacket with electric heater.
- This is the final product; pack ghee in consumer package.
- Use warm detergent solution to clean the complete plant from inside and outside; and wipe out all traces of butter/ghee.
- Clean all pipes/valves.

### 3.3.1  DIRECT CREAM METHOD

Dairy industries with low scale of production of ghee are not preparing butter from cream, instead they directly prepare the ghee from cream separated from milk by centrifugation. Fresh cream, either cultured or washed, is heated to about 115°C in a jacketed ghee kettle of stainless steel, fitted with an agitator, steam control valve, pressure, and temperature gauges, and a movable hollow, stainless steel tube centrally fitted for emptying out the ghee. Alternatively, a separate provision is made for a tilting device on the ghee kettle to decant-off the product. Heating is discontinued as soon as color of ghee turns to golden yellow or light brown, and avoids the burn flavor. One of the disadvantages of the direct cream method is that it requires a long heating time to remove the moisture.

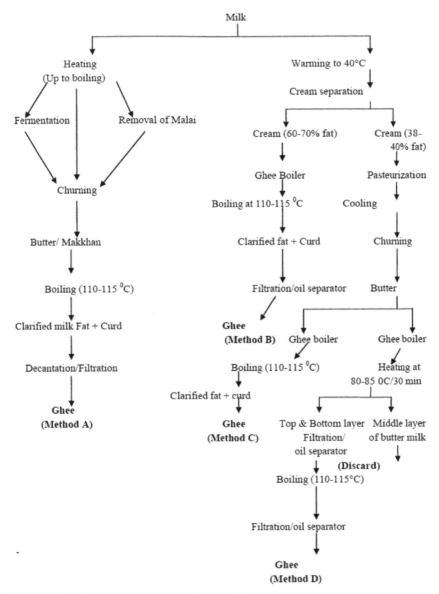

**FIGURE 3.3**   Flow chart to prepare ghee by different methods.

High content of serum solids in the cream is to produce a highly cara-melized flavor in the ghee, and loss of about 4–6% of butterfat in the ghee residue occurs, along with handling operations. There is minimum fat loss and steam consumption in processing of cream or washed cream with about 75–80% fat used for ghee preparing.

### 3.3.2   CREAMERY BUTTER METHOD

This is the most adopted method of many organized dairies where either unsalted creamery butter or white butter, or both are used for ghee making.[7,8] A typical assembly for this method comprises the following equipments:

1.   A cream separator.
2.   Butter churner.
3.   Butter melting outfits.
4.   Steam-jacketed, stainless steel ghee kettle with agitator and process controls.
5.   Ghee filtration devices, such as disc filters or oil clarifier.
6.   Storage tanks for cream, butter, and ghee.
7.   Pumps and pipelines interconnecting these facilities.
8.   Crystallization tanks.
9.   Product filling and packaging lines.

Initially, the butter is melted at 60°C. The molten liquid butter is pumped into the ghee boiler, and the steam pressure valve is opened to release the steam supply to raise the temperature to boil the ghee. The scum, which is collected on the top surface of the product, is removed intermittently with the help of a perforated ladle. After the removal of maximum moisture, the temperature is increased slowly in a controlled way, until the last stage of product is reached. At the stage of the disappearance of effervescence, appearance of finer air bubbles on the surface of fat, and browning of the curd particles takes place, and the typical ghee aroma is also developed, and the final temperature of clarification is adjusted to about 110°C. The filtered ghee is then pumped to the storage tank.

### 3.3.3   PRE-STRATIFICATION METHOD

The pre-stratification procedure consists of keeping the melted butter undisturbed in a ghee boiler at 80–85°C for 30 min to stratify the mass into three separate layers.[12]

### 3.3.4 EFFICIENCY OF DIFFERENT METHODS OF GHEE MAKING

The efficiency of ghee making varies from 80 to 85% for indigenous methods compared to 88–92% for creamery butter method and 90–95% in direct cream method. The energy requirement is highest in indigenous direct cream and lowest in pre-stratification method.[5,6]

## 3.4 PRESERVATION OF GHEE

Storage life of ghee is about 9 months at 21°C. On prolonged storage at ambient temperature, it experiences several defects such as: decrease in nutritive value; destruction of vitamins and carotene; formation of toxic products; loss of attractive color and unsaturated fatty acids; and production of an objectionable flavor.

The durability of ghee can be increased by adopting practices such as: use of antioxidants;[4,9,11] use of good quality raw material (Fig. 3.3); and use of lacquered tin cans.[9]

## KEYWORDS

- CIP cleaning
- continuous ghee making plant
- foaming
- fouling
- furnace oil
- ghee filling machine
- preservation of ghee

## REFERENCES

1. Abichandani, H.; Sarma, S. C.; Bector, B. S. Continuous Ghee Manufacture: An Engineering Solution. *Indian Food Ind.* **1991,** *10* (4), 35–37.

2. Agrawala, S. P.; Prasad, S. A. D.; Nayyar, V. K. Development of Ghee Filling Machine. *Indian Dairyman*. **1980,** *32,* 239–240.
3. Ganguli, N. C.; Jain, M. K. Ghee: Its Chemistry, Processing and Technology. *J. Dairy. Sci*. **1973,** *56,* 19–25.
4. Kuchroo, T. K.; Naranyan, K. M. Preservation of Ghee. *Indian Dairyman*. **1973,** *25,* 405–407.
5. Pandya, A. J.; Singh, J.; Chakraborty, B. K. Energy Consumption of Ghee Making by Direct Cream Method. *Egyptian J. Dairy Sci*. **1987,** *15,* 145–150.
6. Pandya, A. J.; Singh, J.; Chakraborty, B. K. Energy Consumption of Ghee Making by Indigenous Method. *Asian J. Dairy Res*. **1987,** *6,* 21–25.
7. Parekh, J. V. Ghee and Its Technology. *Dairy Technol*. **1978,** *9,* 32–35.
8. Punjrath, J. S. New Developments in Ghee Making. *Indian Dairyman*. **1974,** *26,* 275–278.
9. Rajorhia, G. S. Ghee. In *Encyclopaedia of Food Science, Food Technology & Food Nutrition;* Academic Press: London, 1993; pp 2186–2192.
10. Ramamurthy, M. K.; Narayanan, K. M.; Bhalerao, V. R. Effect of Phospholipids on Keeping Quality of Ghee. *Indian J. Dairy Sci*. **1968,** *21,* 62–63.
11. Rao, C. N.; Rao, B. V. R.; Rao, T.; Rao, G. R. T. M. Shelf-life of Buffalo Ghee Prepared by Different Methods by Addition of Permitted Antioxidants. *Asian J. Dairy Res*. **1984,** *3,* 127–130.
12. Ray, S. C.; Srinivasan, M. R. *Pre-stratification Method of Ghee Making*. ICAR Res. Series No. 8; ICAR, Krishi Bhawan: New Delhi, 1975; p14.

# CHAPTER 4

# TECHNOLOGICAL INTERVENTIONS IN KULFI PRODUCTION: A REVIEW

KUMARESH HALDER[1*], AMIT JUNEJA[1], SUBROTA HATI[2], and CHANDAN SOLANKI[3]

[1]*National Institute of Food Technology Entrepreneurship and Management, Kundli 131028, Sonepat, Haryana, India*

[2]*Dairy Microbiology, SMC College of Dairy Science, Anand Agricultural University (AAU), Anand 388001, Gujarat, India*

[3]*Central Institute of Post-Harvest Engineering and Technology, Ludhiana 141001, Punjab, India*

*Corresponding author. E-mail: kumar.halder@gmail.com*

## CONTENTS

## ABSTRACT

Kulfi is a traditional frozen dairy product of Indian origin prepared by concentrating milk with added sugar, fruits, nuts, flavors, and colors. It is very palatable and nutritious like icecream. Since its inception, the product has been mostly prepared and sold by small-scale street vendors. The physicochemical, microbiological, and sensory qualities of the product vary widely due to different production methods, product formulations, and compositional variation in the ingredients used by the producers. However, because of the huge demand, a few systematic studies were conducted for establishing standard protocol for the product and the production method.

## 4.1  INTRODUCTION

India has emerged as the largest milk producer in the world with an annual production of 146.3 million tons during 2014–2015,[34] accounting for 17% of the total world milk production. It is estimated that about 0.7% of total milk is being converted into icecream, kulfi, and other frozen desserts. Per capita consumption of these frozen desserts in India is about 100 mL/annum, compared to 22 L/annum in the USA.[30] A few food companies have got positions in the Indian market, only based on these frozen desserts. Among these, kulfi (frozen dairy dessert or Indian icecream) is a very popular and widely consumed milk product in India.[12,13]

Kulfi is a traditional Indian frozen dairy product, popular among all age groups, especially during summer months due to its refreshing characteristics, delightful sweet taste, and distinctive cooked flavor. This is mostly popular in northern parts of India. However, the product is also popular in other Asian countries like Pakistan, Bangladesh, Nepal, and Burma (Myanmar). According to the Food Safety and Standards Authority of India (FSSAI) (2006), "ice cream," "kulfi," "chocolate ice cream," or "softy ice cream" means the product is obtained by freezing a pasteurized mix prepared from milk and/or other products derived from milk, with or without the addition of nutritive sweetening agents, fruit and fruit products, eggs and egg products, coffee, cocoa, chocolate, condiments, spices, ginger, and nuts, and it may also contain bakery products such as cake or cookies as a separate layer and/or coating. According to FSSAI Standards,

*"The said product may be frozen hard or frozen to soft consistency; the said product shall have pleasant taste and shall be smell free from off flavor and rancidity. The type of icecream shall be clearly indicated on the label, otherwise standard for icecream shall apply."*

"Frozen Dessert/Frozen Confection" means that the product is obtained by freezing a pasteurized mix prepared with milk fat and/or edible vegetable oils; and fat having a melting point of not more than 37°C in combination and milk protein alone or in combination or vegetable protein products singly or in combination with the addition of nutritive sweetening agents, for example, sugar, dextrose, fructose, liquid glucose, dried liquid glucose, maltodextrin, high maltose, corn syrup, honey, fruit and fruit products, eggs and egg products, coffee, cocoa, chocolate, condiments, spices, ginger, and nuts. The said product may also contain bakery products such as cake or cookies as a separate layer or coating, and it may be frozen hard or frozen to a soft consistency. *"It shall have pleasant taste and flavor free from off flavor and rancidity"*, according to FSSAI

Cow or buffalo milk is traditionally preferred for preparing kulfi. It comes in many flavors, including pistachio, *malai*, mango, cardamom, apple, orange, peanut, avocado, anchovy, and saffron. The composition of kulfi is almost similar to its Western counterpart icecream.[3,16,27,35] Kulfi differs from western icecream in that it is richer in taste, creamier in texture, practically contains no air and it is a solid and dense frozen milk product, whereas ice-creams are whipped with air or produced with a significant overrun. The indigenous process of kulfi mix preparation includes concentration of milk in an open pan and adding sugar, nuts, as well as flavor and with or without adding stabilizer or color into that. The mixture is filled in metal containers of special types, for freezing in the earthenware called *matka*, and then filled with a mixture of ice and salt. Despite the change in modern dairying, this product is still prepared in the unorganized sector. Because of this, very few published literature on the process of manufacturing are available.[5,14,18,26,35]

This product is manufactured locally in small scale because the materials are readily available at low costs as well as the process is simple, and because of that, it has poor microbiological quality, and widely varying chemical and sensory qualities. Despite of all these, this product is having rising demand in the market due to its pleasing sensory attributes, which leads to rise in prices.[15] The involvement of equipment and machinery in the manufacture of kulfi is minimum.[1,25] There is no significant improvement in the traditional method of kulfi making, which is produced mainly by the unorganized sector. Due to nonavailability of a standardized process for its manufacturers, very few organizations are manufacturing it on a commercial scale.[27] However, those dairies have decided to take up the challenges for manufacturing of kulfi,[2,14] so it is expected to have a sea change in present status.[24]

This chapter focuses on the potential and production of kulfi under Indian conditions.

## 4.2  HISTORICAL PERSPECTIVE

The word "kulfi" is derived from the Persian word *qulfi*, for a covered cup, and from the Hindustani word *kulaf* which means a lock or container that has to be unlocked. According to the detailed record of the Mughal emperor Akbar's administration, this product originated during the Mughal period in the 16th century (approximately 500 years ago), and was invented by mixing pistachios, nuts, and saffron with evaporated milk, which was very popular during the same period for preparing various Hindu sweet dishes.[3] The mixture was generally packed in metal cones of special types for freezing in the ice slurry. In *Ain-i-Akbari*, it is also mentioned, the use of saltpeter as well as transportation of Himalayan ice covering miles of distance to warmer areas for refrigeration.

## 4.3  DIVERSIFICATION

Due to the ever-growing demand of kulfi, especially in India by varied health conscious consumers, researchers have been encouraged to develop new types and varieties of kulfi, which will give value added benefits out of consuming kulfi. These include kulfi from admixtures of partially de-oiled groundnut meal and milk/milk powders,[24] sugar free kulfi,[16] sapota pulp admixture kulfi,[31] golden kulfi,[11] etc. Kulfi is also served with *falooda* (vermicelli noodles made from starch). In some places, people make it at home and make their own flavors like rose or any flavor with sugar syrup and other ingredients.

## 4.4 COMPOSITION OF KULFI

According to the Prevention of Food Adulteration (PFA) standards of 1976, it may contain permitted stabilizers and emulsifiers, not exceeding 0.5% by weight. The product shall contain not less than 10% milk fat, 3.5% protein, and 36% total solids, and except the product containing fruits or nuts or both, the content of milk fat shall not be less than 8.0% by weight. Starch may be added to a maximum extent of 5%, under a declaration on the label. The drawback in large-scale production is that PFA regulations do not differentiate between plain icecream and kulfi with regard to its composition. Chemical composition of kulfi has been listed in Indian Standards (ISI 10501-1983).[17]

Several studies on different samples from the market as well as laboratory have been conducted to evaluate the exact composition for producing the optimum quality product. Wide variations in the final chemical composition of kulfi were observed due to variations in the raw materials quality and quantity, and method of production. In most of the cases, total solids, fat, solids not fat (SNF), and protein content were significantly less than the samples prepared in laboratory using regulatory standards, according to the reports by researchers.[21,26,27,35] It has been reported that the product containing 18% milk solids not fat (MSNF) with 3% starch has the best sensory attributes.[5] For industrial production, an ideal composition of kulfi mix was figured out as: fat—11%, MSNF—16%, sugar—15%, and isabgol—0.2%. In most of the cases, market samples fail to meet the minimum legal requirements.

## 4.5 ROLE OF RAW MATERIALS IN FORMULATION OF KULFI

Ingredients required to prepare kulfi may be of dairy as well as nondairy origin. Milk is the major raw material that acts as a source for milk solids (fat and MSNF). Non-dairy materials may include sweetener, stabilizer, emulsifier, color, flavor, and other additives.

### 4.5.1 DAIRY INGREDIENTS

Dairy ingredients, mainly milk and cream have important roles in producing good quality kulfi in relation to body, texture, nutritional, and sensory properties. However, few other dairy ingredients (e.g., sweetened condensed skim milk or partly skimmed milk, evaporated skim milk, and milk powder, etc.) can be used to compensate SNF content in kulfi. Studies on standardization of kulfi mix have revealed that the mixture of equal parts of cow milk and buffalo milk produces the desired quality kulfi. However, a typical kulfi formulation constitutes 9% milk fat, 17% MSNF, 13% sugar, and 1–2% nuts. Therefore, approximately 65% total solids of kulfi are composed only of milk solids.[3] If the mix is formulated, containing 26% milk solids, then it is said to have optimum body, texture, sensory properties, and overall quality. The pH of the mix is 6.4 and surface tension is 46.2 dynes/cm. Therefore, milk is standardized to 3.5% fat and 8.5% SNF for kulfi preparation; then the fat present in final product (7.9%) will not be able to meet the minimum legal requirement of fat (10%).[27] Therefore, the choice of dairy ingredients and the formulation of kulfi mix should be determined based on regulatory standards, desired quality of frozen dessert, marketing strategy, consumer demand, relative price, and availability of ingredients.

### 4.5.2   ROLE OF MILK FAT

Milk fat has a very crucial role in producing good quality kulfi. It contributes to rich creamy flavor, mellowness, smooth texture, fat-soluble vitamins, and provides better melting resistance in the product.[3] It is essential to mix the correct proportion of fat so that the final desired quality product can be produced with minimum legal requirements. It has been reported that the mix containing 7.9% milk fat produced kulfi with the most desirable texture and body.[27] In another study, it was suggested that milk fat content in the range of 13—14% will produce a superior product. The PFA standard for kulfi prescribes a minimum of 10% fat. However, higher amount of fat may lead to produce mix with greater viscosity, which hinders whipability, and contributes to oxidized, rancid and fishy flavor defects in kulfi. Therefore, the amount of fat for preparing kulfi mix should be determined by regulatory standards, which will help to produce kulfi with desired uniform quality.

### 4.5.3   ROLE OF MSNF

MSNF provides valuable proteins, minerals, and vitamins, and improves the body and texture of kulfi. Particularly, milk proteins have a major role for making the body compact and smooth. Higher MSNF (20.5%) causes sandiness defect and cooked flavor with a heavy and soggy body for the product, while lower MSNF (16.2%) produces the product with a weak and crumbly body.[3] More solid level produces kulfi with undesirable texture, which may be due to coarse denatured particles in kulfi, on account of heating and concentration of milk.[16] An experiment claimed that 9.9–12.2% MSNF will produce kulfi with superior body and texture.[35] However, for producing optimum quality kulfi, studies recommended the use of 18.3% MSNF.[27]

Concentrated whole milk is generally preferred as the source of MSNF in kulfi. However, superheated condensed skim or whole milk can substitute the heat-concentrated milk now used by halwais. Properly condensed milk improves whipping ability and binds more free water. Accordingly, less water is available to form crystals during freezing and the smoother texture of kulfi is maintained throughout its shelf-life. Sweetened condensed skim or whole milk may be used as a source of SNF. Added sugar imparts better keeping quality. Due to high osmotic pressure, growth of microorganisms is suppressed. Titrable acidity test should be done for all condensed milk products. The amount, quality, and cost of MSNF affects the commercial production of the product. To eliminate this issue, the recent trend has been to incorporate less expensive nondairy ingredients like starch into kulfi mix. Samples of kulfi prepared with 3.0% starch and 18.0% MSNF showed superior organoleptic properties.[5]

### 4.5.4   NON-DAIRY INGREDIENTS

#### 4.5.4.1   SWEETENERS/SUGAR

Sugar has a very important role in providing a pleasant sweet taste to kulfi. In addition, it increases the total solids. Cane sugar is preferably used for its low cost and high-energy value. Too much sugar impedes the freezing process because of soluble components and excessive sweetness to the product.[3] A satisfactory level of cane sugar in kulfi has been reported as 13[27]–15%.[35] However, other sweetening agents like beet sugar, corn syrup, HFCS, and invert sugar are also used in icecream. Jaggery can also be used as replacement for cane sugar. Jaggery is nutritious as it contains more minerals and is easily available. Jaggery contains more mineral content than cane sugar. It also contains glucose and fructose.

Stevia, the artificial sweetener can be used for this purpose, but due to bitter after taste of stevia up to 50%, sugar replaced stevia and sweetened kulfi is made without affecting the physiochemical and sensory properties of the product.[16] The dietary options of substituting sugar with artificial sweeteners may be especially helpful in the management of ailments like obesity and diabetes. Artificial sweeteners like aspartame and maltodextrin may be used in the preparation of kulfi.[20]

## 4.5.4.2 STABILIZERS AND OTHER ADDITIVES

Stabilizers are mainly added to produce gel-like structure in combination with water, and the gel helps to provide a smooth body and texture. A perfect gel prevents the growth of ice-crystals during freezing as well as at storage, and improves melting resistance in the product. Among the stabilizers used in the manufacture of kulfi and other frozen dairy products are sodium alginate, guar gum, and isabgol husk. Sodium alginate is the salt of alginic acid extracted from brown algae of the class, *phaeophyceae*. Alginates contain D-mannopyranosyluronic acid unit (M) and L-gulopyrasosyluronic acid unit (G). The management intensive grazing (MIG) ratio in the stabilizers differs from source to source and influences the solution properties. Viscosity of alginate solution decreases as the temperature is raised. It is influenced only slightly by pH change in the range of 4–10. The optimum level of sodium alginate recommended for kulfi is 0.15%.[5] Since no overrun is desired in kulfi, it would be necessary to know whether lower or higher percentages of this stabilizer will be ideal. Isabgol husk (*Psyllium* seed husk) is obtained from Plantago sp. that is cultivated in the Western parts of India. It is made up of arabino-xylan units. The potential of isabgol husk as an icecream stabilizer has been evaluated by scientists.[32] Addition of 0.3 and 0.4% isabgol husk imparted good body and texture to icecream. It may be used as a cheap substitute for sodium alginate in kulfi.

Frozen desserts are popular mainly because of their pleasant flavor and refreshing characteristics. The flavoring materials are of three types, namely, natural or true flavor, artificial or imitation flavor, and compound flavors, made by combining natural flavor with artificial *flavor*. The most preferred flavor in kulfi is natural cardamom. Cardamom is the fruit of *Elettaria cardamomum*. The seeds contain 3.0–5.0% of aromatic oil. The mixture of cardamom, pistachio, and almonds may be added in the ratio of 1:2:2 @ of 0.5% by weight in the kulfi mix before freezing to improve the flavor.[22]

## 4.6   TECHNOLOGICAL INTERVENTIONS IN KULFI PRODUCTION

Cow, buffalo, or mixed milk either as such or partially skimmed is concentrated in a large open pan over fire.[4] Raw cow milk standardized to 3.5% fat and 8.5% SNF was concentrated in an open pan jacketed steam heated vessel with constant stirring until the required solids content was attained. Slow heating is preferred to avoid burnt particles in the final product.[27] The involvement of equipment and machinery in the manufacture of kulfi is minimum.

It is suggested to concentrate milk to half of its original volume or in the ratio 2:1. Concentrated milk is cooled, sugar is added and the blend is thoroughly stirred. In some cases, small quantity of khoa or nonfat dried milk is added during boiling of milk. When it has cooled down, *malai* (indigenous cream, crushed nuts, and a flavor (commonly rose or vanilla)) is added. After adding nuts and saffron, the mixture is filled into triangular, conical, or cylindrical moulds of various capacities made of galvanized iron sheets. The moulds are closed on top by placing a small disc over them and the edges are made air-tight with wheat dough. Modern moulds are made up of plastic, generally conical in shape with screw-cap plastic tops. The mix in moulds is frozen in a large earthen vessel, containing a mixture of ice and salt in the ratio of 1:1.[12,27] Vigorous agitation is applied from time to time to improve heat transfer and expedite the freezing process. Agitation of cones apparently whips up some air and kulfi develops a minor overrun. The traditional method of making kulfi in ice-salt mixture by immersing kulfi cones gave the best sensory quality. If the mix is frozen and extruded from an icecream freezer, the quality was quite acceptable. Direct hardening or hand whipping followed by hardening procedures did not produce good kulfi of desired quality. The process flow diagram for kulfi preparation has been described by Salooja.[26,27]

There is no significant improvement in the traditional method of kulfi making, which is produced mainly by the unorganized sector. However, with certain dairies taking up the manufacture of kulfi,[2,14] the scenario is likely to change.[24] Due to non-availability of standardized process for its manufacturers, very few organizations are manufacturing on a commercial scale.[27] In most of the cases, market samples failed to meet the minimum legal requirements.

## 4.7  EFFECTS OF PROCESSING CONDITIONS ON FROZEN DESSERTS

### 4.7.1  HEAT TREATMENT

Pasteurization is an important unit operation to heat every particle of the mix at a predetermined time–temperature combination for killing of all pathogens present and making the kulfi safe for consumption. It performs another noticeable function in proper mixing of all ingredients during the processing. Hence, this improves both flavor and shelf-life of the products. For pasteurization of kulfi mix, the optimum time–temperature combination suggested by Indian Standard (IS: 10501-1983) is 68.5°C for 30 min or 80°C for 25 s. The recent trend is toward higher temperature and lower holding time. Sterilization is not very common in heat treatment for kulfi mix;[36] however, it has been reported that if the mix containing 37–40% total solids is sterilized at 121°C for 15 min then it has good organoleptic scores and its keeping quality was also noticeable.[36]

### 4.7.2  HOMOGENIZATION

Homogenization is mainly done to form a stable emulsion by reducing the size of fat globules preferably to less than 2 microns. This improves the body, color, flavor, palatability, and digestibility of the product. Homogenization helps in imparting smooth texture and body characteristics, besides a rich flavor and greater digestibility of fat,[1] and makes the product glossy and stable. The benefits of homogenization in kulfi making have not been clearly understood.

### 4.7.3  FREEZING AND HARDENING

Freezing of mix is one of the most important unit operations in the manufacture of kulfi. Both quality and palatability of the finished product depends on the freezing technique. The freezing of kulfi may be done in two phases.

In first phase, freezing kulfi in conventional salt/ice freezer and partial freezing in mechanical freezer followed by hardening increased the stiffness of the product and increased melting resistance, compared with the product prepared by direct freezing without agitation. When the kulfi mix is whipped and partially frozen in refrigerator and hardened in salt and ice mixture at −15°C for 3–4 h, it results in a product of good organoleptic quality.[35]

## 4.7.4 PACKAGING

Traditionally, kulfi is packaged in plastic or aluminum moulds, which are partially thawed by dipping in fresh water to remove the kulfi prior to serving. In certain areas, *malai*-kulfi is served as slices of kulfi. Here, kulfi is frozen as a cylindrical block, which is covered by an insulating cloth, and then sliced for serving by vendors. Entrepreneurs may utilize packaging technology used in icecream novelty packaging, using attractive shapes and colors. Kulfi may also be produced on sticks.[3] Kulfi is either sold in an individual cone of triangular, conical, or cylindrical forms of galvanized iron sheets, or in capped plastic moulds. The net weight of market samples of kulfi in cones varied from 95.0 to 107 g. The retail price per cone of kulfi during 1997–1998 ranged from Rs. 10 to 14 (1 US$ = Rs. 60.00). Despite its wide fluctuating properties, kulfi is generally preferred to icecream, because of low cost and detectable sensory attributes. Retort technology has been developed and perfected for packing omega-3 enriched sterilized kulfi samples.[29] Proper packaging material must be used to improve the product quality during storage.

## 4.8   QUALITY CHARACTERISTICS OF KULFI

### 4.8.1   PHYSICOCHEMICAL ASPECTS

The different constituents exert profound influence on the physicochemical properties of kulfi mix. The physicochemical properties like acidity, pH, surface tension, viscosity, and melting quality are altered because of the changes in the relative proportions of the constituents and the processing parameters.[6] Milk proteins and mineral salts have direct influence on the acidity and pH of the mix. Acidity increases and pH of mix deceases with increase in amount of MSNF added to the mix. Sterilization of kulfi mix at 121°C was found to increase the titratable acidity and decrease the pH.[35] Surface tension is a force resulting from an attraction between surface molecules of a liquid that gives surface film like characteristics. The greater the attraction between the molecules, the higher is the surface tension. It decreases with the increase of surface-active agents like emulsifiers, proteins, etc. A decreased surface tension favors faster rate of air incorporation in the mix during the freezing process.[1] Surface tension of kulfi mix was decreased from 50.76 to 45.02 dynes/cm with increasing level of total milk solids from 17 to 29% and this was due to higher concentration of proteins and other organic substances.[26] Viscosity is the internal friction which tends to resist the sliding of one part of the fluid over another. At one time, high viscosity was believed to be important; but for fast freezing in modern equipment, a lower viscosity seems desirable. The average values for viscosity of kulfi mix increases with increase in milk solids due to the higher concentration of MSNF and high fat content leading to partial clumping of fat.[26]

The addition of pre-gelatinized starch and other stabilizers increases the viscosity of mix significantly. Kulfi mix containing 37.1–40.3% total solids had a viscosity range of 28.1–36.6 cP[35] but after sterilization of the same mix at 121°C, the viscosity got increased from 66.53 to 128.53 cP.[35] The rate of melting is dependent on the constituents, total solids, protein content, processing conditions, nature, and concentration of the stabilizer-emulsifier system used in kulfi mix. It has been reported that the melting resistance increases with increase in milk solids.[26] The formation of gel structure by addition of sodium alginate and starch to kulfi mix increases the stability of mix and provides greater resistance toward the melting of kulfi.[5] Melting rate was decreased due to reduction in sugar level or lactose concentration and increased moisture content of kulfi. Therefore, at higher levels of sugar replacement, increase in free moisture content and subsequent increase in large ice crystal formation might be the reason for decreased melting rate.[16]

## 4.8.2 MICROBIOLOGICAL QUALITY REQUIREMENT

Studies conducted on microbiological quality of kulfi[6] were mostly related with the contamination and possible health hazards from that. Pathogenic strains of *Escherichia coli* attracted considerable attention in view of their suspected role in the outbreaks of gastroenteritis in humans and animals.

Presence of pathogenic serotypes in the market kulfi indicated potential contamination for causing serious health hazards. Staphylococci in milk products have drawn considerable attention in recent years as potential food poisoning organisms due to their ability to produce enterotoxins under favorable conditions. The incidence of staphylococci and staphylococcal thermonuclease were also encountered in market samples of kulfi.[8,19,23] Preformed enterotoxins and thermostable deoxyribonuclease present in raw milk due to staphylococcal contamination can survive subsequent heat treatment and are carried over to milk products like *khoya* and kulfi.[7,33] Enterococci, because of their ubiqutious nature, are invariably present in milk and milk products. In view of their ability to resist low temperatures and other unfavorable conditions, these organisms are likely to be present in frozen dairy products, indicating the prevalence of improper sanitary conditions during processing. Among the enterococci isolated from kulfi, *Streptococcus faecalis* var. *liquefaciens* and S*treptococcus faecium* were observed.[10] Thermonuclease positive enterococci, *S. faecium* and *S. faecalis* var. *fecal* have shown that pathogenicity and enterotoxigenicity were also detected in some samples of kulfi.[9] It is now evident that the microbiological quality of raw milk, processing and handling of product during the preparation of kulfi is very important.

### 4.8.3  SENSORY ASPECTS

The sensory qualities of most preferred samples of kulfi by the consumers were described as slightly cooked to caramelized flavor, creamy taste, fine grainy texture, and slight brown color. Addition of cardamom, pistachio, almonds, cashew nuts, etc., adds variety to kulfi. The presence of large-sized ice crystals and coagulated milk particles and fast melt down diminished the product acceptability.

### 4.8.4  SHELF-LIFE

Kulfi is frozen and normally used quickly after it is made. If a large-scale manufacturing operation is undertaken, a shelf-life of 4–8 weeks can be achieved by using technology associated with icecream production. By the judicious selection of a stabilizer emulsifier system, textural defects like coarseness due to ice crystals and grittiness due to lactose crystals can be avoided in stored kulfi. Storage and distribution at a temperature of −20°C (with minimum fluctuations) should maintain the sensory quality of kulfi.[3] Usually, kulfi has a shelf-life of 10 months when kept frozen, but their quality deteriorates due to pathogenic contamination. Sugar present in kulfi itself acts as a preservative by plasmolysis.

## 4.9 VALUE ADDITION

Several attempts have been made for value addition to increase the acceptability of product among the consumers. Studies were conducted to prepare dietetic kulfi and the effect of partial replacement of sugar with stevia on the quality of dietetic kulfi.[16] At higher levels of sugar replacement, there was a significant decrease of specific gravity, melting rate, and significant increase of freezing point, hardness, fat, protein, ash, and moisture content. Retort processing of omega-3 enriched kulfi concentrates to increase the nutritional value as well as shelf-life of the product.[29] Jambhul powder was also used in the development of bioactive components enriched milk kulfi. It was found that the incorporation of 3% of jambhul powder showed value addition in the milk kulfi in terms of enhanced phenolic and antioxidant capacities.[28]

The preparation of kulfi from admixtures of partially de-oiled groundnut meal and milk/milk powders was attempted. This study indicated that an acceptable quality of kulfi could be prepared by replacing up to 50% of milk solids with groundnut solids.[24] The golden kulfi prepared from buffalo milk and safflower milk blend (60:40) was found acceptable by the consumers.[11] The levels of artificial sweeteners were optimized for preparation of sugar free kulfi. It was found that the usage of aspartame and sorbitol, at quadratic level, have a significant effect on the overall acceptability of the product.[20] The protein and fat content were decreased with increase in levels of jaggery and fruit pulp in kulfi and the non-reducing sugar, reducing sugar, iron, ash, and total solid contents were increased in the end product, with increase in levels of both jaggery and fruit pulp.[31]

## 4.10 MECHANIZED PRODUCTION

The production of kulfi is still confined to the unorganized cottage industry or is produced by small-scale entrepreneurs. Very few attempts have been made for mechanized production of the product. Because of delicacy, the demand is growing in the market and in a few places, it is preferred to icecream. Therefore, mechanization is essential to meet that demand and for large-scale commercial production. The production of kulfi on a semi-industrial scale is being carried out in several metropolitan cities in India. In this process, standardized milk is concentrated to about half of the original volume or to PFA standards. The mix is frozen in an icecream freezer with a little overrun (10% or less) and filled into moulds. Moulds are chilled quickly in a brine tank at −20°C for 15–20 min. The frozen product is further hardened for 6 h at −20°C. In some cases, single serve small earthenware *kasora* or *matka* may be used for freezing kulfi.[4] A compact cast-iron freezing unit, which is generally an imported one and consists of a retort connected to a specially designed condenser, is used for freezing the mix. In general, the method of freezing (in a closed system) consists of first heating the retort containing some crude ammonium salts over an open fire, while the condenser is kept immersed in a tub of cold water. This heating goes on for nearly 3 h. Thereafter, the condenser is taken out of the tub and the vessel containing the mix is placed in the annular space in cold water, provided in the condenser and the retort. Some wet cloth or gunny is put on the vessel containing the mix. The freezing takes about 3 h. The frozen product is then taken out of the vessel, wrapped with an insulating material such as paper-and-felt, and sold by chipping out slices with a sharp knife.[12] There is a lot of scope for adapting fully automated endless conveyor belt system, with slots to hold metal cones through freezing liquid/brine mixtures for kulfi production.[4]

## 4.11  CONCLUSIONS

Kulfi is popular throughout India, Pakistan, Bangladesh, Nepal, Burma (Myanmar), and the Middle East, and widely available in Indian restaurants in Australia, Europe, East Asia, and North America. As popularly understood, kulfi has similarities to icecream in appearance and taste; however, it is denser and creamier. Unlike Western icecreams, kulfi is not whipped, resulting in a solid, dense frozen dessert similar to traditional custard-based icecream. Thus, it is sometimes considered a distinct category of frozen dairy-based dessert. However, large variations have been observed due to production in unorganized sector and lack of a standard protocol for production in physicochemical, microbiological qualities, and sensory attributes of kulfi. Yet, now all the organized dairies and food producers are producing kulfi along with other frozen desserts because of its high demand. Due to its higher density, kulfi takes a longer time to melt than its western counterpart and a golden opportunity is waiting for this product to cover a wider market. Technological advances in value addition and standardization of the products have been reported, but a well-established process for mechanized production has been recommended in this chapter for a uniform quality nutritious product produced under hygienic conditions.

## KEYWORDS

- ice cream
- kulfi
- stabilizer
- sweetener
- value addition

## REFERENCES

1. Arbuckle, W. S. *Ice Cream*, 4th ed.; Van Nostrand Reinhold Company: New York, 1986.
2. Aneja, R. P. Traditional Milk Specialties: A Survey. In *Dairy India*; Devarson Stylish Printing Press: New Delhi, India, 1992; pp 268–269.
3. Aneja, R. P.; Mathur, B. N.; Chandan, R. C.; Banerjee, A. K. *Technology of Indian Milk Products*; A Dairy India Publication: New Delhi, India, 2002.
4. Aneja, R. P. East-West Fusion of Dairy Products. In *Dairy India Yearbook*, 6th ed.; S. Gupta, Ed.; A Dairy India Publication: New Delhi, India, 2007; pp 51–53.
5. Ashokraju, A.; Ali, M. P.; Reddy, K. K.; Reddy, C. R.; Rao, M. R. Studies on the Preparation and Quality of Kulfi. *Indian J. Dairy Sci.* **1989,** *1,* 127–129.
6. Bandyopadhyay, A. K.; Mathur, B. N. Indian Milk Products: A Compendium. In: *Dairy India*, 3rd ed.; P. R. Gupta, Ed.; Dairy India: New Delhi, India, 1987; p 1008.
7. Batish, V. K.; Chander, H. Occurrence of *Staphylococcus Aureus* and Their Preformed Enterotoxins in Frozen Dairy Products. *Aust. J. Dairy Technol.* **1987,** *42,* 22–24.
8. Batish, V. K.; Chander, H.; Ranganathan, B. Screening of Milk and Milk Product for Thermonuclease. *J. Food Sci.* **1984,** *49* (4), 1196–1197.

9. Batish, V. K.; Chander, H.; Ranganathan, B. Prevalence of Enterococci in Frozen Dairy Products and Their Pathogenicity. *Food Microbiol.* **1984**, *1*, 269–276.

10. Batish, V. K.; Ranganathan, B. Occurrence of Enterococci in Milk and Milk Products, II: Identification and Characterization of Prevalent Types. *N. Z. J. Dairy Sci. Technol.* **1984**, *19*, 189–196.

11. Bhadakawad, A. D.; Adangale, S. B.; Shinde, D. B.; Mitkari, K. R.; Khating, L. E. Preparation of Golden Kulfi from Buffalo Milk Blended with Safflower Milk. *J. Dairy. Foods Home Sci.* **2009**, *28* (1), 35–38.

12. De, S. *Outlines of Dairy Technology;* Oxford University Press: New Delhi, India, 1980.

13. Elango, A.; Jayalalitha, V.; Pugazhenthi, T. R.; Dhanalakshmi, B. Prevalence of Psychrotrophic Bacteria in Kulfi Sold in Chennai Market. *J. Dairy. Foods Home Sci.* **2010**, *29*, 97–101.

14. Ghodekar, D. R.; Rao, K. S. Microbiology of Kulfi. *Indian Dairyman.* **1982**, *34*, 257–262.

15. Ghosh, J.; Rajorhia, G. S. Process Development for the Manufacture of Instant Kulfi Mix Powder; In *Annual Report*; NDRI: Karnal, 1992; p 66.

16. Giri, A.; Rao, H. G.; V. R. Effect of Partial Replacement of Sugar with Stevia on the Quality of Kulfi. *J. Food Sci. Technol.* **2014**, *51*(8), 1612–1616.

17. *ISI 10501, Specification for Ice-cream.* Indian Standard Institute: Manak Bhavan, New Delhi, India, 1983.

18. Itzerott, G. Notes on Milk and Indigenous Dairy Products in Pakistan. *Dairy Sci. Abstract.* **1960**, *22*, 325–327.

19. Kahlon, S. S.; Grover, N. K. Incidence of Staphylococcai in Milk Products Sampled from Ludhiana. *Indian J. Dairy Sci.* **1984**, *37*, 381–383.

20. Naik, A. P.; Londhe, G. K. Optimization of Levels of Artificial Sweeteners for Preparation of Sugar Free Kulfi. *J. Dairy. Foods Home Sci.* **2011**, *30* (1), 15–24.

21. Natarajan, A. M.; Nambudripad, V. K. N. Microbial Contamination from Ice Cream Containers. *Indian J. Dairy Sci.* **1980**, *33* (4), 500–503.

22. Rao, K. S. Microbiological Quality of Kulfi. M.Sc. Thesis, Kurukshetra University, Kurukshetra, India, 1978.

23. Rao, K. S.; Batish, V. K.; Ghodekar, D. R. Screening Kulfi for Staphylococcal Enterotoxins with Thermonuclease Test. *J. Food Prot.* **2008**, *43*, 49–57.

24. Ramachandran, L.; Sukhminder, S.; Ashwani, R. Preparation of Kulfi from Admixtures of Partially De-oiled Groundnut Meal and Milk/milk Powders. *Nat. Prod. Radiance.* **2005**, *4* (2), 90–96.

25. Rajor, R. B.; Vani, B. A New Approach for Manufacture of Kulfi. *Indian Dairyman.* **1991**, *43*, 256–259.

26. Salooja, M. K. Studies on Standardization of Techniques for Kulfi Production. M.Sc. Thesis, Kurukshetra University, Kurukshetra, India, 1979.

27. Salooja M. K.; Balachandran, R. Studies on the Production of Kulfi, Part–I: The Acceptable Level of Total Milk Solids. *J. Food Sci. Technol.* **1982**, *19* (3), 116–118.

28. Sonawane, K. S.; Arya, S. S.; Gaikwad, S. Use of Jambul Powder in the Development of Bioactive Components Enriched Milk Kulfi. *J. Microbiol. Biotechnol. Food Sci.* **2013**, *2* (6), 2440–2443.

29. Siddharth, M.; Divya, S.; Ramasamy, K. Retort Processing of Omega-3 Enriched Kulfi Concentrates. *Trends Biosci.* **2014,** *7* (22), 3617–3621.
30. Sodhi, R. S. Indian Ice Cream Market: Amul Ice Cream Success Story. *Indian Dairyman* **2004,** *56* (6), 66–68.
31. Ubale, P. J.; Hembade, A. S.; Choudhary, D. M. To Study the Effect of Level of Jaggery and Sapota Pulp on Chemical Quality of Kulfi. *Res. J. Animal Husbandry Dairy Sci.* **2014,** *5* (2), 62–67.
32. Upadhyay, K. G.; Patel A. R.; Vyas, S. H. Evaluation of Isabgol (*Psyllium*) Husk and Gum Acacia as Ice Cream Stabilizer. *Guj. Agric. Univ. Res. J.* **1978,** *4* (1), 45–50.
33. Varadaraj, M. C.; Nambudripad, V. K. N. Carry Over of Preformed Staphylococcal Enterotoxins and Thermostable Dedeoxyribonuclease from Raw Cow Milk to Khoa: A Heat Concentrated Indian Milk Product. *J. Dairy Sci.* **1986,** *69,* 340–343.
34. Department of Animal Husbandry, Dairying & Fisheries, Ministry of Agriculture, GoI, Statistics on Milk Production in India; 2016. www.nddb.org/information/stats/milkprodindia (accessed on November 7, 2016).
35. Yerriswamy, K.; Atmaram, K.; Natarajan, A. M.; Anatakrishnan, C. P. Quality of Kulfi Manufactured by Different Methods. *Cheiron* **1983,** *121* (3), 130–135.
36. Yerriswamy K.; Atmaram K.; Natrajan, A. M.; Anatakrisnshn, C. P. Preparation of Sterilized Kulfi Mix. *Cherion* **1984,** *13* (5), 223–226.

# CHAPTER 5

# HYBRID TECHNOLOGY FOR THE PASTEURIZATION OF MILK

SHILPI SINGH[1], JAYEETA MITRA[2*], and VISWANATHA ANGADI[3]

[1]*Food Process Engineering, Department of Agricultural and Food Engineering, Indian Institute of Technology, Kharagpur 721302, India*

[2]*Department of Agricultural and Food Engineering, Indian Institute of Technology, Kharagpur 721302, India*

[3]*Department of Food Microbiology, Agriculture College Hassan, University of Agricultural Sciences (UAS), Bangalore 573225, India*

[*]*Corresponding author. E-mail: jayeeta12@gmail.com*

## CONTENTS

## ABSTRACT

In this chapter, authors discussed thermal and non-thermal pasteurization of milk with mechanisms through which these technologies act on microorganisms. Authors also presented research advances on hybrid technologies in order to give additive or synergistic effects on overall milk quality. The detailed advantages of hybrid technologies for milk pasteurization were presented. Since these are emerging technologies and studies are still going on to validate these technologies, researchers can come with more precise hybrid technologies with their respective effects.

## 5.1   INTRODUCTION

Pasteurization is nothing but a mild heat treatment given to a particular food. The two main objectives of pasteurization are, to kill the pathogenic microbes from foods and hence prevent different types of diseases, and to remove the spoilage microbes hence enhancing their shelf-life. Louis Pasteur in 1857 observed that the milk souring could be delayed by giving heat treatment to milk at about 122–142°F (50.0–61.0°C), although the agents that were responsible for the souring/spoilage of milk were not firmly established until later. After a lot of studies on the properties of milk and microbes, it was established that the primary cause of these problems are microorganisms.

By 1895, it was recognized that milk can be made safe for consumption by heating it to a predetermined temperature for a definite time. By the year 1927, North and Park had investigated a wide range of time–temperature combinations (130°F for 60 min up to 212°F for 10 s) for inactivating *tubercle bacilli* that was known to be the index microorganisms at that time and responsible for the disease called tuberculosis.

Heat treatment is predominantly used for the preservation of milk and other liquid food materials. For safe and shelf stable milk and milk products, removal of spoilage and pathogenic microbes is very essential. Although, thermal processing can be considered as an ideal food preservation technique, as it provides a solution to a wide range of problems related to food industry. However, in some foods, the high thermo-tolerance of some thermoduric microorganisms (mainly bacterial spores) and enzymes necessitate the application of high temperature and longer residence time, which somehow imparts adverse effects on the nutritional and sensory properties of milk. Therefore, alternative solution to thermal processing in order to remove pathogenic and spoilage microorganisms have been identified by the dairy industry and non-thermal technology provides very attractive solution to this problem.

Non-thermal treatment signifies processing of milk without application of heat. However, practically non-thermal processing can be considered as either processing of milk without heat application or processing of milk with much lower heat treatment as compared to the treatment that is normally attained in thermal processing. Therefore, it ensures a minimal degradation of milk quality. However, non-thermal technologies should have characteristics to improve the nutritional values and milk quality, and to ensure an equivalent or, preferably, an increase in the safety levels with respect to other preservation techniques that they replace. When we combine non-thermal methods with other processing techniques, we should ensure that:

- It reduces the severity of the non-thermal treatments to attain same level of inactivation.
- Enhance the lethality effects of the particular treatment, and/or
- Prevent multiplication of survivors that have undergone the treatments.

This chapter discusses overview of the commercial thermal, non-thermal technologies, and hybrid technologies for milk pasteurization.

## 5.2 THERMAL PASTEURIZATION OF MILK

Based on exhaustive experimentation, a definite set of temperature and time combinations has been set for the pasteurization of milk, that is, 63°C for 30 min for batch low temperature long time (LTLT) pasteurization, 72°C for 15 s for continuous high temperature short time (HTST) pasteurization and 83°C for 3 s for ultra-high temperature (UHT) pasteurization. Table 5.1 provides a clear view about how do we reach for these temperature and time combinations of pasteurization.

Initially, the target microorganism was mycobacterium tuberculosis. But in recent discoveries, it has been replaced by *Coxiella burnetii*, which is responsible for Q-fever (Table 5.1). However, it has been already stated that pasteurization is responsible for killing only the pathogenic microbes, not all. Those microbes, which survive heat treatment as severe as pasteurization, are termed as thermoduric; and those, which have an ability to tolerate even a more harsh treatment as high as 80–100°C for 30 min, are known as spore formers. Hence, the pasteurized milk requires quick chilling right after heating as a resistance for the thermoduric microorganism's growth.

**TABLE 5.1**   Temperature and Time Combination for Different Actions in Milk.

| Requirement | Temperature °C | |
|---|---|---|
| | For 30 min | For 15 s |
| Cream line reduction | 63.3 | 72.22 |
| Kill *Coxiella burnetii* | 62.2 | 71.66 |
| Pasteurization requirement | 63.0 | 72.00 |
| Phosphatase inactivation | 61.1 | 71.11 |

Nowadays, HTST continuous process is being commercially used by the industry. The continuous operations has several advantages over the batch process, namely: possible automation, higher and quick heating, and cooling rates, less exposure time to heat, regeneration, facilitation of online monitoring, and much more. Also, it is affordable if we do not consider the establishment costs. Scales of operations on continuous heat exchangers range between 500 and 50,000 lph, with experimental models as down as to 50 lph.

In addition to this, continuous processing introduces some additional complexities in the industry, like attaining: flow control, pressure control, accurate holding time, and accurate temperature but all these problems can very well be resolved by dairy industries. Schematic diagram (Fig. 5.1) shows the industrial view of a milk pasteurization unit. However, consumer demand toward minimally processed milk/food products is increasing day-by-day as it witnessed minimum harm to the sensory properties and minimum degradation to the nutritional properties of milk.[36] Therefore, we should adopt non-thermal methods of milk pasteurization.

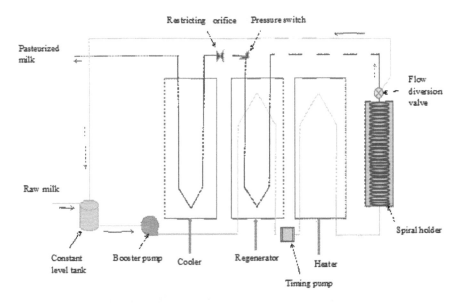

Yellow color lines = flow of raw milk
Brown color lines = flow of pasteurized milk

**FIGURE 5.1**   Schematic diagram of pasteurization process.

## 5.3   NON-THERMAL PROCESSING OF MILK

### 5.3.1   HIGH PRESSURE TREATMENT OF MILK

High pressure processing (HPP) is the application of considerable amount of pressure in order to rupture the cells of the microorganisms and thus inactivate them. As HPP does not involve any type of heat treatment to milk, hence all drawbacks of thermal processing of milk are overcome. Also, HPP treated milk is superior in terms of organoleptic properties compared to thermal processing. Le Chatelier's principle and the Isostatic principle govern the operation of HPP. Le Chatelier's principle indicates that a system retains its equilibrium condition as and when it encounters any disturbance.

Isostatic principle states that the hydrostatic pressure process is volume–independent; therefore, pressure can be applied instantaneously and uniformly throughout a sample, and pressure gradients do not exist so that the size and geometry of the product is irrelevant. This is an advantage for HPP as compared to thermal-processed food, in which the heat treated food products may be over heated at the surface or an insufficient heating may take place at the bulk attracting the growth of microorganisms.

Disadvantages of HPP are: the food materials having large amount of air will disrupt in their shape and size because of the high pressure employed on it making it unsuitable. However, it is not true in case of milk as milk is a liquid. Treatment at 400 MPa for 15 min or 500 MPa for 3 min was equivalent to pasteurization for a shelf-life of 10 days stored at 10°C.[35]

## 5.3.2  PULSE ELECTRIC FIELD TREATMENT FOR PASTEURIZATION OF MILK

The high intensity electric field treatment is being used to induce electroporation, a phenomenon that is used to introduce damage to the microbes DNA by applying numerous perforations in the microbial membranes. Advances in the severity of the treatment and different combinations of treatments have led to the introduction of pulsed electric field (PEF) in dairy industry.[37] During PEF treatment of milk, there are two electrodes placed at a distance that can vary from 0.1 to 1.0 cm in a closed chamber and milk is allowed to pass through between those electrodes. Short pulses of duration of 1–10 µs are generated by a high voltage (5–20 kV) pulse generator. Those pulses heat the milk for fraction of seconds and induce pores in the membrane of microbes. Also, there is a cooling system provided to prevent the growth of thermoduric microbes and to lessen the thermal effects. If the pores formed are small with respect to the surface area of membrane, they will heal themselves in some time. The relationship between trans membrane potential and radius of cells[23] is given below:

$$u(t) = 1.5\, rE, \qquad\qquad (5.1)$$

where $u(t)$ = trans membrane potential in the direction of the applied field ($V$); $r$ = radius of the cell (µm); and $E$ = applied electric field strength (kVmm$^{-1}$).

Pathogens, such as *Escherichia coli, Staphylococcus aureus, Pseudomonas fluorescens, Listeria monocytogenes*, or *Listeria innocua* were inoculated in milk that was treated by PEF processing. There was a significant reduction in pathogen levels between 2 and more than 5 log cycles by applying energy in the range from 100 to 550 kJ/L. Critical factors affecting the effectiveness of PEF processing to inactivate the microbes are:

- The process parameters: electrical field strength, pulse duration, pulse polarity, and treatment time;
- The microorganism itself (physiological state): size and shape of the microbial cell, cell wall characteristics for Gram-positive vs. Gram-negative bacteria, inoculum size, and natural flora vs. inoculation tests; and
- Milk composition: electrical conductivity and fat content.

### 5.3.3  USE OF CARBON DIOXIDE FOR MILK PROCESSING

Apart from high and low temperature treatments, there are about 60 other types of food preservation techniques: to remove/reduce microbial growth in milk and milk products; to preserve the organoleptic and nutritional qualities of a food; and to extend the shelf-life. For each type of food product, the right combination of hurdles (Fig. 5.2) with the right intensities can ensure removal of all pathogens rendering the particular food harmless and safe for consumption, while at the same time challenging its economic viability and consumer preferences.

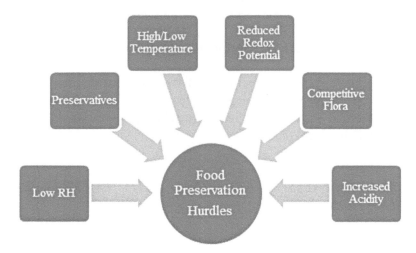

**FIGURE 5.2**   Most popular hurdles during food processing in industry.

Carbon dioxide ($CO_2$) is an alternative hurdle technology that can reduce the microbial load and have very good potential to prevent growth of spoilage microbes and to extend the shelf-life of foods with a better efficacy. High pressure $CO_2$ has come up with a good solution for dairy industry to treat with same effectiveness at low to moderate temperatures. Also, it ensures better functionality of milk as compared to thermal pasteurization of milk.[41] The effectiveness of $CO_2$ as an antimicrobial hurdle needs more investigation as there can be synergistic effect when combined with other preservation hurdles mentioned above and may carry an antagonistic effect. For example, $CO_2$ with reduced water activity has antagonistic effects on food.

## 5.3.4   HIGH POWER ULTRASOUND PROCESSING OF MILK

The ultrasound has rapidly emerged as a mild non-thermal processing having a potential of replacing or assisting many conventional food processing applications, such as emulsification, homogenization, mixing, milling, extraction, pasteurization, filtration, moisture removal for drying,[30] and crystallization. For majority of food applications, ultrasound frequency ranges from 20 to 40 kHz. Mode of action of ultrasound for inactivating microbes is due to three mechanisms: cavitation, free radical formation, and localized heating of milk hence, attaining a reduction of microbes equal to pasteurization.

## 5.4   USE OF HYBRID TECHNOLOGY FOR PASTEURIZATION

Hybrid technology implies inclusion of two or more thermal and non-thermal technologies for a synergistic effect that will give the same lethal effect as thermal pasteurization; and it provides better organoleptic properties and higher level of safety.

## 5.4.1   HIGH PRESSURE COMBINED WITH TEMPERATURE REGULATION

It has been investigated that microbial cells can be readily ruptured in low viscosity media; however, high viscosity media limit the use of high pressure homogenization (HPH). When it comes to the resistance of microbes, then Gram-positive bacteria are more resistant to HPH-induced inactivation than Gram-negative bacteria.[11,22] Therefore, researchers suggest that HPH combination with heat can act as an alternative for the pasteurization and homogenization of milk. A combination of pressure greater than 150 MPa and temperature greater than 40°C resulted in microbial inactivation that can be somewhat equivalent to those achieved during HTST pasteurization processes.[20,32,33,38] When stored at refrigerated temperature, shelf-life of such HPH milk is almost equivalent to that of traditionally HTST pasteurized/homogenized milk. In addition, the HPH-induced disintegration of fat molecules and reduction in size of fat globule help in decreasing the extent of creaminess in milk.

## 5.4.2 PULSE ELECTRIC FIELD COMBINED WITH TEMPERATURE REGULATION

The efficiency of PEF processing can be significantly increased by hybridization of the system with some amount of heat, that is, combining the effects of the electric voltage pulses with temperature in order to reduce microbial load.[15] Temperature may decrease the trans membrane potential of the membrane making it more susceptible to the electroporation and electrical breakdown. The reactions of the pathogens of *Pseudomonas* spp. in low fat milk that has followed the PEF treatment at different processing temperatures (15–55°C) have been investigated and it was concluded that a certain amount of heat with PEF treatment can act synergistically.[8] Log reduction of 3.2 to more than 5.0 was identified by increasing temperature up to 55°C. A combination of PEF and thermal treatment were used to inactivate *L. innocua* in whole milk.[19] The treatment of low fat milk and whole milk at 60°C and 46.5 kV/cm indicated a higher degree of inactivation of the natural micro flora (mesophilic and psychrophilic bacteria). A significant reduction of 3.6 log cycle was observed following the combined treatment.[4]

The properties of whole milk and skim milk and its coagulation with PEF treatment were studied. Firmness of curd was increased and time taken for clotting was reduced indicating an enhancement in the coagulation properties of milk.[15] The effects of PEF processed (20–30 kV/cm, 20–50°C) and heat processed (63°C, 30 min) milk for coagulation properties of milk by rennet enzyme were compared. As expected, the curd firmness made out of PEF treated milk was low. However, the decrease in the firmness was less compared to that of the thermally treated sample.[44] Figure 5.3 shows an example of an industrial pasteurization plant combined with PEF and thermal processing for milk.

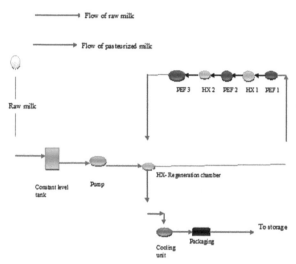

**FIGURE 5.3**   Schematic of a commercial PEF and thermal system for pasteurization of milk.

Similar to conventional unit, it has a regeneration unit where raw milk is heated to some extend with the milk leaving processing chamber. The milk is then passed through the system designed with three processing units. The number of processing units varies with requirement of different foods. After this, milk leaves the processing chambers, and the hot milk goes to regeneration chamber and then to the final cooling section where it is cooled to a temperature of 7°C. The preheating temperature may vary from 50 to 65°C. The regeneration system has the advantage of being cost as well as energy-effective.

### 5.4.3   ULTRASOUND COMBINED WITH HEAT (THERMO-SONICATION)

When we use thermo-sonication, the additional temperature that we impose on milk serve as an additive to the localized heating produced by sonication, and leads to a sufficiently high temperature required for cell lysis and inactivation of microbes.[12,14]

When microbes were treated with a pulsed sonication at 24 kHz and at ambient temperatures of 30–35°C, it induced mechanical damage to cells and membrane.[18] It has been proved that lactose in milk has a protective function associated with it toward microbial cell. Sonication of 24 kHz in combination with a temperature 63°C was used to inactivate *L. innocua* and to reduce the mesophilic bacterial count in raw whole milk by 0.69 log after 10 min and a reduction as high as 5.3 log cycle after 30 min, which ultimately resulted in greater shelf-life of milk.[3] However, when we treat UHT milk instead of raw milk with the same sonication parameters (63°C and 24 kHz), it resulted in prevention of mesophilic bacterial growth higher than 2 log when stored at ambient and refrigerated temperature for 16 days.[2] All these advantages make thermo-sonication an attractive option to extend the shelf-life of pasteurized milk. In addition to this, effect of thermo-sonication against some of the pathogens like *B. subtilis, S. aureus, Salmonella typhimurium,* coliforms were studied and total plate counts were performed for milk. However, literature on exact combination of heat and ultrasound is very limited and it needs more investigation to come up with accurate combination and their effects on microbes.

### 5.4.4 HIGH PRESSURE AND TEMPERATURE COMBINED WITH SUPERCRITICAL CARBON DIOXIDE

Under a sufficiently mixed condition, the combined technique of HP, temperature along with supercritical $CO_2$ has shown significant lethal effect on microbes in milk pasteurization. High pressure, temperature, and time with proper concentration of super-critical carbon dioxide ($sCO_2$) treatment has potential of inactivating the microbial population with a better efficiency than those achieved by commercial HTST pasteurization.[41] When skim milk is treated at 15 MPa and 35–40°C with $CO_2$ concentration to be one-third that of milk, it gets a shelf-life of 35 days or more.[10] Since milk, viscosity has an effect on degree of inactivation, under same processing conditions the microbial count in whole milk is supposed to be higher than skim milk.[13,25] Some of the processing parameters must be increased that includes higher temperature or pressure, longer holding time or higher concentration of $CO_2$ to maintain the same degree of pasteurization to whole milk as that of skimmed milk.

The killing efficiency can be tremendously increased with the mode of mixing of $CO_2$ in milk. In a proper well-mixed sample at 6 MPa and 45°C, a 6 log reduction of *Listeria* in skim milk was observed in 1 h while it took 16 h for the same reduction to take place for unmixed sample. This is because diffusion of $CO_2$ to the interior of milk takes time.

As milk fat has adverse effects on the on $sCO_2$ pasteurization process, we can increase the efficacy of the system either with some of thermal treatment with $sCO_2$ treated milk or by removing the milk cream from raw milk before the process. In the second method, we can treat the cream that we have removed from milk with UHT system and skim milk with $sCO_2$ separately and then introduce the cream at the end of the process to get whole milk. But this process will again question the cost economics of the whole system. The $sCO_2$ treatment can be regarded as a *mild* pasteurization process for skim milk/whole milk under ideal processing parameters. Also, it maintains the nutritional and sensory properties with a greater shelf-life.[10] Spores in milk that are difficult to inactivate by other processes can be easily inactivated with the treatment of $sCO_2$.[41] Also, $sCO_2$ in combination with PEF (25 kV/cm and 20 pulses) treatment at 40°C with a pressure 20 MPa for 24 h were studied, and it resulted in the reduction of *Bacillus* spores count by three log cycles.

### 5.4.5   LOW PRESSURE AND TEMPERATURE REGULATION COMBINED WITH CARBON DIOXIDE

Low-pressure carbonation ($CO_2$) can inhibit the growth of psychotropic bacteria without any loss of sensory as well as nutritional quality; thus ensuring a better shelf-life. Apart from mesophilic microbes, psychrophilic microbes that were not destroyed in pasteurization process starts multiplication when stored at chilled condition and are responsible for the spoilage of milk.

Shelf-life of milk can be extended by 25–200% by the addition of very low concentration of $CO_2$ (@ 1.8–3.2 mM) prior to HTST pasteurization.[21] Some of the spores like *B. cereus* are commonly found in milk and milk-based products[1] and are extremely thermoduric[27,31] and are threats for extended shelf-life. Growths of *B. cereus* spores are responsible for initiating proteolysis and lipolysis of the milk. These changes may incorporate off-odors and may cause coagulation of milk, as well as may produce toxins that can cause either emetic or diarrheal food-borne illness to humans. Under such conditions, carbonation offers advantages. For example, growth of *B. cereus* or *Clostridium botulinum* spores is inhibited by the addition of low-pressure $CO_2$, over long-term storage of 35–60 days.

The killing efficiency of microbes, that are susceptible to these mentioned combinations, increases exponentially with temperature for $CO_2$ treated raw milk and hence the pasteurization temperature reduces considerably. The inactivation of microbes susceptible to the presence of $CO_2$ is proportional to the concentration of $CO_2$ that was used to treat raw milk. However, sensitivity of different microbial species may vary to the combination of carbonation and pasteurization. For example, *P. fluorescens* inoculated to milk pretreated with 36 mM of $CO_2$ shows a 5 log reduction after heating for 35 min at 50°C, but *B. cereus* spores are barely affected after 25 min at 89°C, with less than a 0.3 log improvement compared to heat-treated controls only.[26] The point at which $CO_2$ is injected into the pasteurization line may also have an effect on the microbial kill. For example, $CO_2$ could be added during come-up time in the regeneration section, just ahead of the holding tube, or under pressure during homogenization. Table 5.2 indicates different pasteurization techniques with processing parameters and their advantages.

**TABLE 5.2** Different Pasteurization Techniques with Processing Parameters and Advantages.

| Pasteurization technique | Processing parameters | Advantages | Reference |
|---|---|---|---|
| $CO_2$ + LP + Heat | 1.8–3.2 mM $CO_2$ at 72°C for 15 s | Extend the shelf-life of the milk by 3 times, prevent heat-resistant spores from germinating | [16] |
| High pressure | 500 MPa for 3 min. | Smaller fat globules and homogenized | [25] |
| HPH + Heat | >150 MPa at >40°C | Reduced heat treatment and decreased extent to creaming | [15] |
| HTST | 72°C for15 s | Commercially available | |
| PEF + Heat | 20–30 kV/cm at 20–50°C | reduced number of pulses (energy consumption), enhancing the coagulation properties of milk | [33] |
| Pulse electric field (PEF) | 25–37 kV/cm at 15–60°C | Low energy usage | [27] |

**TABLE 5.2** *(Continued)*

| Pasteurization technique | Processing parameters | Advantages | Reference |
|---|---|---|---|
| sCO$_2$ + HP + Heat | 15 MPa and 35–40°C with a CO$_2$/milk ratio of 0.33 | Significantly increase the shelf-life, increased killing efficacy, reduced heat treatment | [7] |
| Thermo-sonication | 63°C and 24 kHz | Lower mesophilic count in ambient temperature | [2] |

## 5.4.6 BACTERIOCINS FOR MILK PASTEURIZATION TECHNOLOGY

Commonly employed preservation techniques are presented in Figure 5.2. Bacteriocins can be used in combination with other treatments to inactivate microbial cells and get the similar effect as that of thermal pasteurization.[39] However, a detailed study is needed to optimize the use in dairy industry and pasteurization process. Table 5.3 indicates recent studies on effects of combining bacteriocins with other technologies on the shelf-life.

Nisin is one of the most commonly used bacteriocins in food industry as this is the only one commercially approved for use in food industry. Therefore, most of the studies have been done on nisin. However, the antimicrobial documented effects within dairy industry, its efficiency in combination with additional hurdles need more investigation for its use in foods as they may incorporate antagonistic effects and other environment related problems. In a study, it was found that nisin showed a reduced potential when used in combination with other preservatives, that is, NaCl.

### 5.4.6.1 BACTERIOCINS COMBINED WITH HEAT TREATMENTS

The presence of nisin in milk has an antimicrobial effect associated with it; therefore, it reduces the severity of heat treatments required for inactivation of certain population of microbes, when combined with thermal treatment.

**TABLE 5.3** Different Combinations of Bacteriocins with Different Conventional and Emerging Technologies.

| Bacteriocin | Additional hurdles | Target microorganism | Dairy product | Reference |
|---|---|---|---|---|
| Nisin (328 IU/mL) | High pressure (654 MPa), Heat (74°C) | *Clostridium perfringens* spores | UHT milk | [16] |
| Nisin (400 IU/mL) | Thymol (0.08 mg/mL) | *L. monocytogenes* | UHT milk | [43] |
| Nisin (250 IU/mL) | Polylactic acid (1 g) | *L. monocytogenes* | Skimmed milk | [24] |
| Nisin (0.75 µg/mL) | Endolysin LysH5 (15 U/mL) | *S. aureus* | Milk | [17] |
| Enterocin AS-48 (28 AU/mL) | High intensity pulsed electric field (HIPEF, 800 µs) | *S. aureus* | Milk | [39] |
| Nisin (20 IU/mL) | High intensity pulsed electric field (HIPEF, 800 µs) | *S. aureus* | Milk | [39] |
| Nisin (500 IU/mL) | High pressure (500 MPa) | *B. cereus* spores, *B. subtilis* | Reconstituted skimmed milk | [5] |
| Nisin (300 IU/mL) | HIPEF (1200 µs) | *S. aureus* | Milk | [39] |
| Nisin (0.04 µg/mL) | PEF (16.7 kV/cm) Carvacrol (1.2 mM) | *B. cereus* | Milk | [34] |
| Nisin (75 and 150 IU/mL) | Heat treatment (117°C, 2 s) | Spore-forming bacteria (Bacilli) | Milk | [42] |
| Nisin (100 or 200 IU/mL) | Lactoperoxidase | *L. monocytogenes* | Skimmed milk | [6] |
| Lacticin 3147 (10,000 or 15,000 AU/mL) | HHP (150, 275, 400, or 800 MPa) | *S. aureus* *L. innocua* | Milk and whey | [29] |
| Nisin (10 or 100 IU/mL) | PEF (30, 40, and 50 kV/cm) | *L. innocua* | Skimmed milk | [7] |
| Nisin (2000 or 4000 IU/mL) | Heat (97, 100, 103, and 130°C) | *B. cereus, Bacillus stearothermophilus* | Skimmed milk | [40] |
| Nisin (10 or 100 IU/mL) | Lactoperoxidase (0.2 or 0.8 ABTSU/mL) | *L. monocytogenes* | UHT Skimmed milk | [45] |

When whole milk is treated with 25 IU/mL of nisin at a temperature of 54°C for 16 min, a 3 log10 reduction of *L. monocytogenes* was observed while it took 77 min to achieve same reduction in absence of nisin which indicates a considerable reduction in time. Also the effect was investigated on *L. monocytogenes* in cheese milk, and it was found that adding 25 or 50 IU/mL of nisin reduces the intensity of heat treatment.[28] Apart from this, it was noted that heat resistance of *L. monocytogenes* was reduced in presence of nisin provided that it was stored at a refrigerated temperature of 4°C before thermal treatment was carried out. A reduced heat treated milk (RHT, 117°C, 2 s) with combination of nisin (75 IU/mL or 150 IU/mL) resulted in tremendously high shelf-life of 150 days stored at ambient temperature of 30°C, whereas in most of the cases, the RHT-milk started showing signs of spoilage only after 15 days of time. Furthermore, when nisin-RHT milk was stored at low temperatures as low as 10 or 20°C, it did not show any type of microbial activity up to one year. Also, RHT-nisin milk showed a more positive response than UHT milk in terms of sensory evaluation.[40,42] Synergistic effects for spore destruction in milk were investigated when it was treated with nisin (4000 IU/mL) and heat. It has emerged that the RHT and the nisin treated milk might be more costly and energy effective technology for pasteurization of milk and may give an improved product with greater shelf-life.[40]

### 5.4.6.2  *BACTERIOCINS COMBINED WITH NON-THERMAL TECHNOLOGY*

Bacteriocins have been successfully used with various non-thermal technologies that include high pressure treatment (HP), PEF, ultra-sonication, and irradiation. However, different methods have different mode of action for inactivation and hence have different after effects of the processes. Some methods enhance the killing efficacy by reducing the inactivation time of pathogens while others prove to be very efficient in reducing the mesophilic and psychrophilic counts hence increasing the shelf-life after pasteurization. The complete representation of the Bacteriocins combined with different technologies and their effects on a particular microbes and the whole process[9] are discussed below.

Probabilistic shelf-life estimation for several stand alone as well as hybrid milk pasteurization technologies have been plotted in Figure 5.4. However, variation may be because of non-uniform reduction of the initial microbial count through processing.

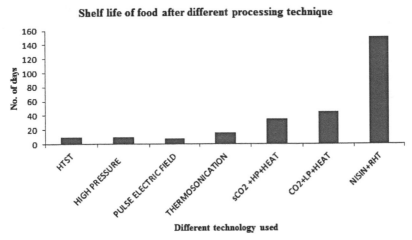

**FIGURE 5.4**    Effect of different technologies on shelf-life of milk.

From Figure 5.4, it can be inferred that nisin with the aid of reduced heat treatment is the best technique, if shelf-life is important. Also it provides a higher cost economy compared to other processes. Use of $CO_2$ can act as an alternative to milk processing in most of the cases.

## 5.5   CONCLUSIONS

There are non-thermal technologies like pulse light, irradiation, ultraviolet treatment, etc., that can be used in combination with other technologies for the processing of milk and milk products. This hybrid technology can act as a potential area of research with multiple benefits like extended shelf-life, reduced energy cost, reduced heat treatment, better organoleptic, and sensory properties, etc. However, extensive research is needed to validate these hybrid technologies. Also, research data on cost economics of these processes is not enough. Although hybrid technologies has yet to find widespread acceptance in the dairy industry, yet these can have potential areas of research as the consumer demand for minimally processed food is increasing every day.

Apart from microbial inactivation, these non-thermal and hybrid technologies show different reactions on different milk constituents and parameters like viscosity, milk fat, milk fat globules, milk proteins, milk salt, milk enzymes, etc.

Although success of hybrid technologies has not been well documented, yet there is an urgent need of further investigation as reported work was carried out in laboratory scale only. Also we need to make sure that the synergistic and additive effects are not altered with change in the food environment. Apart from this, most of these technologies need to be validated and there is a lack of data on the properties that are needed for proper design of equipments.

## KEYWORDS

- **bacteriocins**
- **carbonation**
- **electroporation**
- **homogenization**
- **hurdle technology**
- **hybridization**
- **inactivation**
- **index sonication**
- **transmembrane potential**
- **thermoduric**
- **thermo-sonication**

## REFERENCES

1. Bartoszewicz, M.; Hansen, B. M.; Swiecicka, I. The Members of the *Bacillus cereus* Group are Commonly Present Contaminants of Fresh and Heat-Treated Milk. *Food Microbiol.* **2008,** *25* (4), 588–596.
2. Bermúdez-Aguirre, D.; Barbosa-Cánovas, G. V. Study of Butter Fat Content in Milk on the Inactivation of *Listeria innocua* Atcc 51742 by Thermo-Sonication. *Innov. Food Sci. Emerg. Technol.* **2008,** *9* (2), 176–185.
3. Bermúdez-Aguirre, D.; Corradini, M. G.; Mawson, R.; Barbosa-Cánovas, G. V. Modeling the Inactivation of *Listeria innocua* in Raw Whole Milk Treated Under Thermo-Sonication. *Innov. Food Sci. Emerg. Technol.* **2009,** *10* (2), 172–178.
4. Bermúdez-Aguirre, D.; Fernández, S.; Esquivel, H.; Dunne, P. C.; Barbosa-Cánovas, G. V. Milk Processed by Pulsed Electric Fields: Evaluation of Microbial Quality,

Physicochemical Characteristics, and Selected Nutrients at Different Storage Conditions. *J. Food Sci.* **2011**, *76* (5), S289–S299.

5. Black, E.; Linton, M.; Mccall, R.; Curran, W.; Fitzgerald, G.; Kelly, A.; Patterson, M. The Combined Effects of High Pressure and Nisin on Germination and Inactivation of *Bacillus spores* in Milk. *J. Appl. Microbiol.* **2008**, *105* (1), 78–87.

6. Boussouel, N.; Mathieu, F.; Benoit, V.; Linder, M.; Revol-Junelles, A. M.; Millière, J. B. Response Surface Methodology, an Approach to Predict the Effects of a Lactoperoxidase System, Nisin, Alone or in Combination, on *Listeria monocytogenes* in Skim Milk. *J. Appl. Microbiol.* **1999**, *86* (4), 642–652.

7. Calderón-Miranda, M. L.; Barbosa-Cánovas, G. V.; Swanson, B. G. Inactivation of *Listeria innocua* in Skim Milk by Pulsed Electric Fields and Nisin. *Int. J. Food Microbiol.* **1999**, *51* (1), 19–30.

8. Craven, H.; Swiergon, P.; Ng, S.; Midgely, J.; Versteeg, C.; Coventry, M.; Wan, J. Evaluation of Pulsed Electric Field and Minimal Heat Treatments for Inactivation of Pseudomonads and Enhancement of Milk Shelf-Life. *Innov. Food Sci. Emerg. Technol.* **2008**, *9* (2), 211–216.

9. Mishra, V. K.; Ramchandran, L. *Opportunities for the Dairy Industry: Emerging Dairy Processing Technologies;* John Wiley & Sons Ltd: Chichester, UK, 2015; p 240.

10. Di Giacomo, G.; Taglieri, L.; Carozza, P. Pasteurization and Sterilization of Milk by Supercritical Carbon Dioxide Treatment. In:*Proceedings 9th International Symposium on Supercritical Fluids*, Archachon, France; 18–20 May, 2009; Vol. 2; p C87.

11. Donsì, F.; Ferrari, G.; Lenza, E.; Maresca, P. Main Factors Regulating Microbial Inactivation by High-Pressure Homogenization: Operating Parameters and Scale of Operation. *Chem. Eng. Sci.* **2009**, *64* (3), 520–532.

12. Drakopoulou, S.; Terzakis, S.; Fountoulakis, M.; Mantzavinos, D.; Manios T. Ultrasound-Induced Inactivation of Gram-Negative and Gram-Positive Bacteria in Secondary Treated Municipal Wastewater. *Ultrason. Sonochem.* **2009**, *16* (5), 629–634.

13. Erkmen, O. Effects of High-Pressure Carbon Dioxide on *Escherichia coli* in Nutrient Broth and Milk. *Int. J. Food Microbiol.* **2001**, *65* (1), 131–135.

14. Feng, H.; Yang, W.; Hielscher, T. Power Ultrasound. *Food Sci. Technol. Int.* **2008**, *14* (5), 433–436.

15. Floury, J.; Grosset, N.; Leconte, N.; Pasco, M.; Madec, M. N.; Jeantet, R. Continuous Raw Skim Milk Processing by Pulsed Electric Field at Non-Lethal Temperature: Effect on Microbial Inactivation and Functional Properties. *Le Lait.* **2006**, *86* (1), 43–57.

16. Gao, Y.; Qiu, W.; Wu, D.; Fu. Q. Assessment of *Clostridium perfringens* Spore Response to High Hydrostatic Pressure and Heat with Nisin. *Appl. Biochem. Biotechnol.* **2011**, *164* (7), 1083–1095.

17. García, P.; Martínez, B.; Rodríguez, L.; Rodríguez, A. Synergy Between the Phage Endolysin Lysh5 and Nisin to Kill *Staphylococcus aureus* in Pasteurized Milk. *Int. J. Food Microbiol.* **2010**, *141* (3), 151–155.

18. Gera, N.; Doores, S. Kinetics and Mechanism of Bacterial Inactivation by Ultrasound Waves and Sonoprotective Effect of Milk Components. *J. Food Sci.* **2011**, *76* (2), M111–M119.

19. Guerrero-Beltrán, J. A.; Sepulveda, D. R.; Góngora-Nieto, M. M.; Swanson, B.; Barbosa-Cánovas, G. V. Milk Thermization by Pulsed Electric Fields (PEF) and Electrically Induced Heat. *J. Food Eng.* **2010**, *100* (1), 56–60.

20. Hayes, M. G.; Fox, P. F.; Kelly, A. L. Potential Applications of High Pressure Homogenisation in Processing of Liquid Milk. *J. Dairy Res.* **2005,** 72 (01), 25–33.

21. Hotchkiss, J. H.; Chen, J. H.; Lawless, H. T. Combined Effects of Carbon Dioxide Addition and Barrier Films on Microbial and Sensory Changes in Pasteurized Milk. *J. Dairy Sci.* **1999,** *82* (4), 690–695.

22. Huppertz, T. Homogenization of Milk: High-Pressure Homogenizers. In *Encyclopedia of Dairy Sciences*; Academic Press: San Diego, CA, 2011; p 7.

23. Jeyamkondan, S.; Jayas, D.; Holley, R. Pulsed Electric Field Processing of Foods: A Review. *J. Food Prot.* **1999,** *62* (9), 1088–1096.

24. Jin, T. Inactivation of Listeria Monocytogenes in Skim Milk and Liquid Egg White by Antimicrobial Bottle Coating with Polylactic Acid and Nisin. *J. Food Sci.* **2010,** *75* (2), M83–M88.

25. Lin, H. M.; Cao, N.; Chen, L. F. Antimicrobial Effect of Pressurized Carbon Dioxide on *Listeria monocytogenes. J. Food Sci.* **1994,** *59* (3), 657–659.

26. Loss, C; Hotchkiss, J. Effect of Dissolved Carbon Dioxide on Thermal Inactivation of Microorganisms in Milk. *J. Food Prot.* **2002,** *65* (12), 1924–1929.

27. Luu-Thi, H.; Grauwet, T.; Vervoort, L.; Hendrickx, M.; Michiels, C. W. Kinetic Study of Bacillus Cereus Spore Inactivation by High Pressure High Temperature Treatment. *Innov. Food Sci. Emerg. Technol.* **2014,** *26,* 12–17.

28. Maisnier-Patin, S.; Tatini, S.; Richard, J. Combined Effect of Nisin and Moderate Heat on Destruction of *Listeria monocytogenes* in Milk. *Le Lait.* **1995,** *75* (1), 81–91.

29. Morgan, S.; Ross, R.; Beresford, T.; Hill, C. Combination of Hydrostatic Pressure and Lacticin 3147 Causes Increased Killing of Staphylococcus and Listeria. *J. Appl. Microbiol.* **2000,** *88* (3), 414–420.

30. Mulet, A.; Carcel, J.; Sanjuan, N.; Bon, J. New Food Drying Technologies—Use of Ultrasound. *Food Sci. Technol. Int.* **2003,** *9* (3), 215–221.

31. Novak, J. S.; Call, J.; Tomasula, P.; Luchansky, J. B. An Assessment of Pasteurization Treatment of Water, Media, and Milk with Respect to Bacillus Spores. *J. Food Prot.* **2005,** *68* (4), 751–757.

32. Pereda, J.; Ferragut, V.; Quevedo, J.; Guamis, B.; Trujillo, A. Effects of Ultra-High Pressure Homogenization on Microbial and Physicochemical Shelf Life of Milk. *J. Dairy Sci.* **2007,** *90* (3), 1081–1093.

33. Picart, L.; Thiebaud, M.; Rene, M.; Guiraud, J. P.; Cheftel, J. C.; Dumay, E. Effects of High Pressure Homogenisation of Raw Bovine Milk on Alkaline Phosphatase and Microbial Inactivation. A Comparison with Continuous Short-Time Thermal Treatments. *J. Dairy Res.* **2006,** *73* (04), 454–463.

34. Pol, I. E.; Mastwijk, H. C.; Slump, R. A.; Popa, M. E.; Smid, E. J. Influence of Food Matrix on Inactivation of *Bacillus cereus* by Combinations of Nisin, Pulsed Electric Field Treatment, and Carvacrol. *J. Food Prot.* **2001,** *64* (7), 1012–1018.

35. Rademacher, B.; Pfeiffer, B.; Kessler, H. Inactivation of Microorganisms and Enzymes in Pressure-Treated Raw Milk. *Spec. Publ. Royal Soc. Chem.* **1998,** *222,* 145–151.

36. Ross, A. I. V.; Griffiths, M. W.; Mittal, G. S.; Deeth, H. C. Combining Non-Thermal Technologies to Control Foodborne Microorganisms. *Int. J. Food Microbiol.* **2003,** *89* (2–3), 125–138.

37. Shamsi, K.; Versteeg, C.; Sherkat, F.; Wan, J. Alkaline Phosphatase and Microbial Inactivation by Pulsed Electric Field in Bovine Milk. *Innov. Food Sci. Emerg. Technol.* **2008,** *9* (2), 217–223.

38. Smiddy, M. A.; Martin, J. E.; Huppertz, T.; Kelly, A. L. Microbial Shelf-Life of High-Pressure-Homogenised Milk. *Int. Dairy J.* **2007,** *17* (1), 29–32.

39. Sobrino-López, A.; Martín-Belloso, O. Use of Nisin and Other Bacteriocins for Preservation of Dairy Products. *Int. Dairy J.* **2008,** *18* (4), 329–343.

40. Wandling, L.; Sheldon, B.; Foegeding, P. Nisin in Milk Sensitizes *Bacillus spores* to Heat and Prevents Recovery of Survivors. *J. Food Prot.* **1999,** *62* (5), 492–498.

41. Werner, B.; Hotchkiss, J. Continuous Flow Non-Thermal Co 2 Processing: The Lethal Effects of Subcritical and Supercritical $CO_2$ on Total Microbial Populations and Bacterial Spores in Raw Milk. *J. Dairy Sci.* **2006,** *89* (3), 872–881.

42. Wirjantoro, T.; Lewis, M.; Grandison, A.; Williams, G.; Delves-Broughton, J. The Effect of Nisin on the Keeping Quality of Reduced Heat-Treated Milks. *J. Food Prot.* **2001,** *64* (2), 213–219.

43. Xiao, D.; Davidson, P. M.; Zhong, Q. Spray-Dried Zein Capsules with Coencapsulated Nisin and Thymol as Antimicrobial Delivery System for Enhanced Antilisterial Properties. *J. Agric. Food Chem.* **2011,** *59* (13), 7393–7404.

44. Yu, L.; Ngadi, M.; Raghavan, G. Effect of Temperature and Pulsed Electric Field Treatment on Rennet Coagulation Properties of Milk. *J. Food Eng.* **2009,** *95* (1), 115–118.

45. Zapico, P.; Medina, M.; Gaya, P.; Nuñez, M. Synergistic Effect of Nisin and the Lactoperoxidase System on *Listeria monocytogenes* in Skim Milk. *Int. J. Food Microbiol.* **1998,** *40* (1), 35–42.

# PART II
# Drying Techniques in the Dairy Industry

# CHAPTER 6

# CLASSIFICATION OF DRIED MILK PRODUCTS

RACHNA SEHRAWAT[1], PRABHAT KUMAR NEMA[1],
PRAMOD KUMAR[2], and ANIT KUMAR[1*]

[1]*Department of Food Engineering, National Institute of Food Technology Entrepreneurship and Management, Sonipat 131028, Haryana, India*

[2]*Dairy Chemistry Division, National Dairy Research Institute (NDRI), Karnal 132001, Haryana, India*

*Corresponding author. E-mail: aks.kumar6@gmail.com*

## CONTENTS

## ABSTRACT

Milk powder can be prepared using spray drying or drum drying of milk. The spray dried milk powder retains better physical and nutritional qualities. The hygiene during the manufacturing process is very important to avoid any unintentional contamination and growth of microorganism. Moisture content is an important parameter and should be kept very low to avoid any microbial or enzymatic deterioration. The recent trends in powders include instant milk powders. The industries have become more sophisticated and automated. In spray drying process, single particle drying is getting attention of researchers and industrialists. Also, Industries are keeping drying chamber externally in open air. It saves building cost as well as avoid explosion.

## 6.1   INTRODUCTION

Milk is nature's perfect food; it only lacks iron, copper, and vitamin C but is highly recommended for nutritionists for building healthy body. Some people are unable to digest lactose sugar present in milk but can easily digest milk-fermented products. For example, in yoghurt, lactose is converted into lactic acid by lactic acid bacteria (LAB). Milk (cow or buffalo) has very short shelf-life so it is a perishable commodity and highly susceptible for microbial contamination. It is a rich source of fat (4–6%), proteins (3–4%), carbohydrates, vitamins, and minerals. Also, it has very high water moisture content (> 80%). In order to increase shelf-life, milk processing is required. The shelf-life can be increased by pasteurization, sterilization, and ultra high temperature treatment. The non-thermal techniques (like pulse electric field, high pressure processing (HPP), irradiation, and ultra-sonication) can also enhance the shelf-life. Apart from these treatments, milk can be processed into different products like fermented products, heat desiccated products, clarified butter fat products, frozen products, and dried products.[2] However, drying is one of the oldest techniques to achieve highest shelf-life.

Drying of milk is carried out to remove the moisture present, with the help of media (air and steam) to convert it into powder having low water activity so that microbes cannot grow. Drying of milk powder started in the 20th century and now it is one of the rapidly growing industries. The objectives of drying milk are to increase the shelf-life, eliminate microbial contamination, convert into powder so that transportations cost can be reduced, and it would be easy to handle, also save storage space.

Moreover, the surplus milk is converted to milk powder form during flush season, so that it can be used during lean season to meet the demand of urban population. Other benefits are: to maintains properties similar to that of raw milk on reconstitution; and to provide as raw material for different product formulation like tea, coffee, and chocolates. Milk product powders are being used for reconstitution into milk, as base or filler for sweets, as ingredients in icecream, mayonnaise, bakery, and confectionary industry. Milk powders exhibit good foaming and emulsifying properties. They form gel as pH approaches toward the isoelectric point (4.6) due to reduction of charges on milk proteins but neutral milk do not form gel.

This chapter focuses on classification of dried milk products.

## 6.2   MANUFACTURING PROCESS FOR DRYING OF MILK

### 6.2.1   SPRAY DRIED MILK POWDER

The underlying principle of spray drying of milk is conversion of preheated and concentrated (45–55% total solids) milk into droplets by an atomizer (also known as atomization). When these spherical droplets come into contact with the heating medium steam/air, they get dried and get collected at the bottom of drying chamber. Due to increase in surface area of milk in form of droplets, they get instantaneously dried into particulate or powder up to 3–4% moisture content.[3] Spray dryer is most widely used for large scale drying of milk. Factors governing the operation includes: temperature of heated air, volume, drying chamber design, and velocity of heated air. Spray dried milk powder has particles size between 10 and 200 μm generally with smooth surface and particle of regular and spherical shapes.[9] They have higher reconstitutability than the milk by drum drying. In Figure 6.1, line diagram for spray drier is shown. Spray driers can be classified as:

- Based on atomization of milk: pressure spray, compressed air, and centrifugal disc.
- Based on method of heating air: by direct gas or fuel, indirect by heat exchangers or coil. Based on position of drying chamber: horizontal and vertical.
- Based on number of chambers for drying: one and two.

- Based on method of heat transfer: convection and radiation.
- Based on pressure atmospheric: vacuum.
- Based on direction of air-flow: co-current, counter-current, and parallel.

Benefits of spray drying include more solubility, superior flavor, and better appearance than drum dried milk powder. It is economical method of drying if quantity of raw material is large. Limitation of spray drier plant is that it involves large capital investment in design and building.[8]

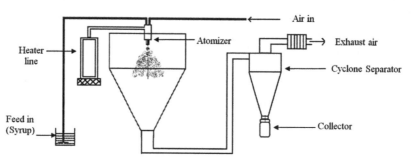

**FIGURE 6.1**   Line diagram for spray drier.

## 6.2.2   DRUM DRIED MILK POWDER

Drum dryer setup involves a supportive mount over which one or more hollow cylinders are kept. Doctor blades are also provided to remove the material from drum surface. The milk is concentrated first (30% total solids) before it is uniformly spread over the outer surface of slowly rotating drum. It is internally heated by steam/air (120–170°C) to reduce the moisture content generally less than 5% on wet basis. Auxiliary feed rollers are also provided to spread the milk uniformly. Dried film formed over the surface of drum is removed by scrappers/knife/doctor blades before completion of one revolution. The dried film sheet obtained is grounded to form powder. Parameters that govern the process of drum drying are: wetting properties, surface tension, viscosity of feed material, steam pressure/hot air temperature, film thickness, and rotational speed. Heat consumption in the operation ranges from 2000 to 3000 kJ/kg of moisture removed. Figure 6.2 indicates the generalized diagram for manufacturing milk.

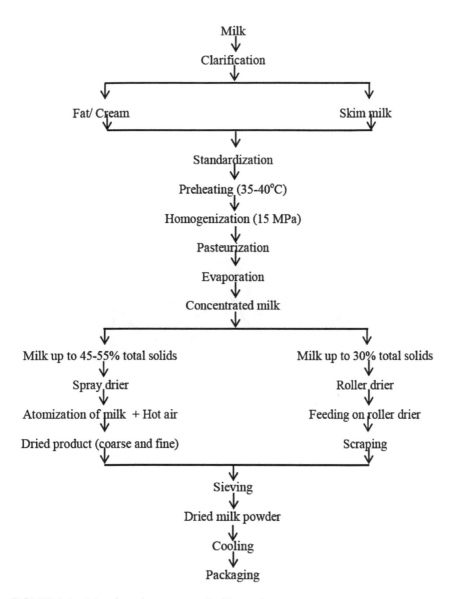

**FIGURE 6.2**   Manufacturing process of milk powder.

Drum drier can be classified based on number of hollow drums (single, twin drums); based on operating pressure (atmospheric, vacuum); directions of drum rotation (twin, double); material of construction (stainless steel, cast iron, steel, alloy, chrome or nickel plated steel, cast iron); and method of feed (top, spray, splash, and bottom feed).

They involve low operating and less capital investment compared to spray driers. They occupy less floor area and is economical to handle small quantity of milk. Powder has good porosity. They suffer from limitation of cooked flavor due to adhering of milk on surface of drum or roller. Powdered milk also has low solubility compared to spray dried powder. Schematic diagram for drum drier is shown in Figure 6.3.

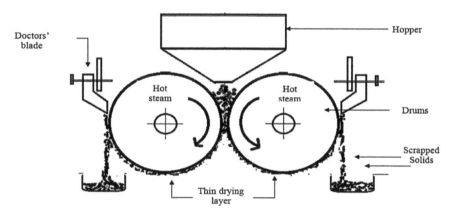

**FIGURE 6.3**   Schematic diagram for drum drier.

## 6.2.3   FREEZE DRIED

In this system, the milk is first frozen below 18°C to solidify it and then sublimation is carried out. Freezing is done to freeze out the water and centrifuging. In second step, heat is added to remove the water by sublimation which involves converting moisture into vapor phase without passing through liquid phase. Vacuum is also maintained in drying chamber. It is carried out at very low temperature under low pressure (known as cold process). The setup for freeze drier consists of vacuum gauge, detachable manifolds, condenser, and vacuum pump (Fig. 6.4). The freeze-drying is also known as lyophilization. Commercial application of freeze-drying is obsolete for drying of milk.

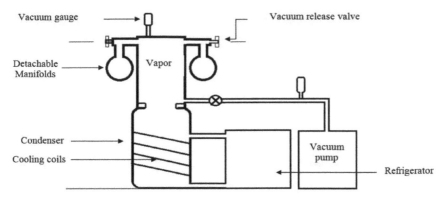

**FIGURE 6.4**   Freeze drier.

## 6.3   CLASSIFICATION OF DIFFERENT DRIED MILK PRODUCTS

Different types of dried milk products are: whole milk powder, skim milk powder, whey powder, butter milk powder, icecream mix powder, infant milk powder, shrikhand powder, chhana powder, cheese powder, khoa powder, and malted milk powder. Flow diagram for preparation of different milk powders is shown in Figure 6.5. American Dry Milk Institute have listed standards for dried milk.[7]

### 6.3.1   WHOLE MILK POWDER

According to the Prevention of Food Adulteration (PFA) Act of 1954, dried milk powder source can be buffalo or cow milk or from standardized milk. Ingredients like citric acid, calcium chloride, and sodium salts of polyphosphoric acids, orthophosphoric acid, and sodium citrate may be added but not exceeding by 0.3% by weight of finished products. The method of drying should be declared on label. Moisture and fat content of whole milk powder should not be more than 5% and not less than 26%, respectively. Solubility index should not be more than 2 and 15 for spray dried and drum dried powder, respectively. The total acidity expressed as lactic acid should not be more than 1.2%. The coliform count and standard plate count may not exceed 90/g and 50,000/g, respectively.[7]

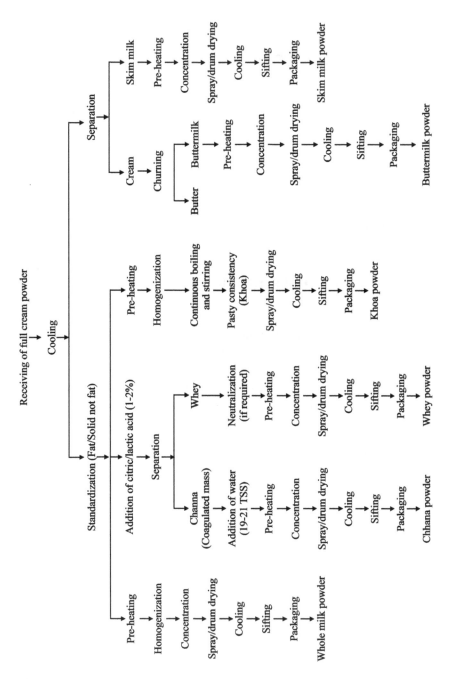

**FIGURE 6.5** Flow diagram for production of dried milk powder.

In the whole milk powder, main hurdle is fat decomposition and flavor deterioration due to high fat content so they are packaged under low oxygen content. To achieve extended shelf-life, gas packing is most effective method. This powder can be used in many ways like for reconstitution into milk, for preparing tea and coffee, for bakery and confectionary products, and also in baby food formulations.

## 6.3.2 SKIM MILK POWDER

According to the PFA Act of 1954, skim milk dried powder source can be buffalo or cow milk or from standardized milk. Ingredients like citric acid, calcium chloride, and sodium salts of polyphosphoric acids and orthophosphoric acid, and sodium citrate may be added but not exceeding by 0.3% by weight of finished products. It is not necessary to mention them on label. The method of drying should be declared on label. Moisture and fat content of whole milk powder should not be more than 5% and not more than 1.5%, respectively. Solubility index should not be more than 2 and 15 for spray dried and drum dried powder, respectively. The total acidity expressed as lactic acid should not be more than 1.5%. The coliform count and standard plate count may not exceed 90/g and 50,000/g, respectively. It is available in market in glass jars, metal cans, and multilayer paper bags with polyethylene inner liner (50–200 μm). It is used in tea, coffee, breads, biscuits, rolls, icecream, Indian sweets, and candy manufacturing.

## 6.3.3 WHEY POWDER

It is yellowish liquid (rich in proteins and low fat) remaining after curdling of milk, which is strained and then dried. Whey is also called by-product of cheese/paneer production. It contains high amount of lactose sugar, which offers difficulty while heating as it becomes sticky. Moreover, during storage it also causes caking due to its hygroscopic nature. If whey is acidic in nature it is neutralized first and then condensed, cooled, seeded with lactose for its crystallization up to 24 h followed by spray drying. Because, it contains hygroscopic lactose sugar, therefore the bags with polythene liner are used for packaging. Sweet whey is utilized in the preparation of processed cheese products, mayonnaise, frozen desserts, bakery, and confectionary products.

### 6.3.4   BUTTERMILK POWDER

It is a slightly sour liquid left after churning of butter. It may be prepared from high acid, sour, and sweet buttermilk. They are rich source of proteins, lactose sugar and minerals, and acid powder also contains high amount of lactic acid. The method of manufacturing involves preheating (32–49°C) to reclaim the fat and then condensing it by evaporation till 40–45% total solids are achieved and then is spray dried. Sour buttermilk can be drum dried at atmospheric conditions if it is to be used for animal and poultry feed as solubility is not of much concern. Some of technical hurdles during drum drying of sour or high acid buttermilk are: unmanageable smear on the drums, high lactose content in dried powder readily absorb moisture from surrounding and becomes sticky to form cakes. These can be sorted out by neutralization of buttermilk before drying, packaging in dry environment, and moisture-proof packaging. They are generally packed in kraft paper bags with plastic liners. They have shelf-life more than a year at room temperature. It can be used to enhance nutritional value of breads, reconstituted beverage and, for animal feed.

### 6.3.5   ICE CREAM MIX POWDER

First of all icecream mix is prepared considering the fat percentage, solid not fat, sugar, emulsifier, stabilizers, and total solids after standardization. The mix is then preheated, homogenized (in first and second stage), again heated after homogenization, and then dried. Dried mix has a shelf-life of 6–12 months if container has 2% oxygen level at 10–15°C. Some of defects observed are browning, staleness, and oxidation.

### 6.3.6   INFANT MILK POWDER

The need to produce the infant milk powder is to avoid malnutrition problems among children. Although mother's milk is a complete food, yet to avoid malnutrition it is another substitute. It is prepared using milk with addition of carbohydrates (cane sugar, dextrose, dextrin, maltose, and lactose), iron salts, and vitamins.

## 6.3.7  CHEESE POWDER

Aged/medium aged cheese are taken and then grounded. Further slurry is prepared by addition of water. Stabilizers and color, sodium chloride may be added. It is then heated, homogenized and then dried, and cooled. It is used for reconstituting into cheese, in preparation of cheese sauce, mayonnaise.

## 6.3.8  SRIKHAND POWDER

Srikhand is prepared from lactic fermented curd after straining it and kneaded with ground sugar to a buttery mass. It is prepared from whole milk having sweetish-sour taste with semi-soft texture. It is consumed directly as sweet dish. The manufacturing process includes developing the srikhand base, which involves preparation of lactic acid fermented curd that is broken and kept in muslin cloth and whey is removed for 8–10 h. This solid mass is also known as chakka. After kneading with sugar (@18%) thoroughly, water (20–25% of mix) is added to it. The slurry obtained is then homogenized and then dried. On gas packaging, the dried product in a hermetically sealed container extends the shelf-life to more than three months.

## 6.3.9  CHHANA POWDER

Milk solids obtained by citric/lactic acid coagulation of milk and removing the whey by straining is known as chhana. Since it is deteriorated within 1–2 days so, in order to extend the shelf-life, drying is carried out. It has high amount of fat and protein. For preparing chhana, cow milk is preferred to achieve the desirable soft texture. The steps in the preparation involve: heating of milk (70°C), coagulation with lactic/citric acid (1–2%) till whey is clear, straining of coagulated mass and whey, cooling of coagulated mass, dilution of total soluble solids in range 19–21% by addition of water, and a slurry is obtained which is heated and finally dried into powder. It has shelf-life of 2–4 months. It is most commonly used as filler for preparation of sweets.

## 6.3.10   KHOA POWDER

It is a heat-desiccated product prepared by concentrating the milk with continuous stirring and scrapping till a semi solid consistency is achieved. To prepare the powder, khoa is first diluted with addition of water till 16–18% total solids are obtained. Then preheated and passed through micro-pulverizer and then dried. It is utilized as filler or base for sweets. Dried powder has storage life more than three months at room temperature.

## 6.3.11   MALTED POWDER

Malted powder is prepared by adding barely malt and wheat flour to the whole milk. Barely malt and wheat flour (2.5:1) is mixed with whole milk, which is heated and then condensed and further drying is carried out. It is used to prepare beverage and also utilized in confectionary.

## 6.4   CHARACTERISTICS OF DRIED MILK PRODUCTS

## 6.4.1   STRUCTURE

Dried milk powder has dual structure primary (dispersion) and secondary (agglomeration). The primary structure comprises powder particles from milk which are in dispersed phase and small amount of moisture and air cells.[5] The secondary structure constitutes of bulk of solid particles, which are closely packed, surrounding the air. The whole particles are not in contact of gas/air. It is formed by joining of primary particles. It is affected by shape, size, and uniformity of the particles.

Lactose present in dried milk is generally present in the non-crystalline state and is hygroscopic. It absorbs moisture readily under humid conditions and solidifies which is known as caking. Heat treatment while drying induces reversible denaturation of protein followed by irreversible coagulation by destabilizing the milk protein. Fat in the dried milk is either surrounded by protein membrane or as free fat. Free fat is literally incorrect. Therefore, called free fat is responsible for greasiness of reconstituted milk as it exists in free-state and generally on the surface of particles. This free state of fat is mainly responsible for promoting oxidation, which can be avoided by homogenization of milk.

## 6.4.2   PARTICLE SIZE

It depends on the manufacturing techniques, operating conditions, and milk characteristics. Agglomerated milk powder has irregular shape and is large in size. The mean particle size (median value of cumulative distribution) ranges from 85 to 230 µm for regular SMP and is 250 µm for fat-filled milk powder, respectively.[5]

## 6.4.3   DENSITY/SPECIFIC VOLUME

Particle density of materials includes vacuoles and is influenced by occluded air (pore soace). Particle density of pressure spray dried milk is higher compared to centrifugal spray dried milk because the entrapped air is higher in centrifugal spray dried milk. Particle density of skim milk powder varies from 900 to 1400 kgm$^{-3}$.

Bulk density refers to weight per unit volume (gmL$^{-1}$). Bulk density is higher of spray dried milk powder than drum dried powder.[10] High bulk density is desirable to decrease the volume. Also, it is higher of pressure spray dried milk than centrifugal spray dried milk. Bulk density is little more of skim milk powder than whole milk powder. Bulk density is reduced to 40–60% by agglomeration in drum-dried powder. Agglomeration is a process during which primary particles are joined together to form bigger porous secondary particles.[5] Bulk density can be increased by removing the occluded air, which can be achieved by pre-concentration (higher the concentration lower will be the entrapped air). Another way to increase the bulk density is by increasing the total solids of the concentrated milk.

True density of particle material excludes the vacuoles. Particle density of whole dried milk is 1300 kgm$^{-3}$, dried skim milk is 1480 kgm$^{-3}$, and of whey dried powder is 1560 kgm$^{-3}$.

Fraction of void spaces over total volume is known as porosity that ranges from 0.4 to 0.75 for dried milk powders.

## 6.4.4   INSTANT PROPERTIES/RECONSTITUTABILITY

To enhance the reconstitutability of milk products powder, they are made instant soluble. Instantization is mainly applied to non-fat dried milk powders, that is, skim milk powder. However, fat rich dried products can also be instantized. These properties depend upon dispersibility, wettability, sinkability, and flowability.

### 6.4.5 DISPERSIBILITY

It is the ability of the individual particles to disperse or separate into individual particles, when reconstituted into water.[4] High temperature treatment of milk containing higher total solids causes irreversible denaturation of proteins to a greater extent. Due to this denaturation, a stable dispersion is not formed by the denatured casein during reconstitution. Proteins are more affected in the drum drying process as compared to the spray drying. Even in spray drying if milk particles are exposed to high temperature and for longer duration, it will enhance the degree of irreversible protein denaturation percentage. Traditional method of indicating denaturation index is whey protein nitrogen index (WPNI). Other indicators for determining extent of denaturation are cysteine number, casein number, heat number, thiol number, furosine content, and sulfhydryl content.[1] Larger particles are generally considered good for dispersibility. Dispersibility of skim milk powder and whole milk powder is more than 90 and 85%, respectively. High dispersible powder exhibits good wettability.

### 6.4.6 WETTABILITY

On interaction of powder and liquid, ability of a powder to absorb water on the surface, wettability refers to be wetted, and to penetrate the surface of still liquid.[5] Lumping of powder on reconstitution indicates poor wettability. Contact angle between water and powder is also correlated with the wettability; higher contact angle indicates less wettability. Lactose being hygroscopic, forms small contact angle indicating good wettability. Agglomeration and absence of fine particles favors wettability. Agglomeration provides more space for penetration of water into interstices. It is also influenced by fat dispersion and fat with low melting point gave better wettability.

### 6.4.7   FREE-FLOWINGNESS

It is the ability of a powder to flow when poured under standardized conditions over the surface as heap. It is estimated by measuring angle made by free flowing powder with the surface known as angle of repose. The angle will be smaller if powder is free flowing. Powder is called sticky if it is not free flowing. This is also improved by different parameters like spherical shape of particles, surface smoothness, agglomeration, and high particle density. Different free-flowing agents (calcium silicates and sodium-aluminum silicate addition) help in improving free-flow ability of powder. It is influenced by moisture content and ratio of solid not fat to fat.

### 6.4.8   SINKABILITY

It is the ability of powder particles to overcome the surface tension of water and sink into the water, after passing through the surface. It is the amount of the powder that sinks per unit time per/over unit surface area. Agglomeration enhances sinkability by increasing the total/aggregate weight.

### 6.4.9   INSOLUBILITY INDEX (II)

On dissolving the powder and liquid/water, the fraction of components that do not get dissolved under standardized conditions of temperature, concentration, time and intensity of stirring; is referred as insolubility index. Solubility is an important parameter and is desirable to form a stable emulsion. The fraction of the milk powder, which is insoluble, is related to heat coagulation of proteins or fat components. It is dependent on temperature, and duration of drying; pre-concentration of milk. Spray-dried whole milk powder, partially skimmed milk powder, and whole milk powder have insolubility index greater than 99%. The solubility of roller dried milk and tray dried cheese powder are about 85 and 91%, respectively.[6]

### 6.4.10   FLAVOR

Flavor of milk should be pleasant, sweet, clean, and rich. Burnt/cooked flavor should be avoided. Burnt flavor mainly develops due to high heat given during preheating possibly during concentration. Another reason for development of cooked flavor is due to heating of fat and Maillard reaction, because during heating methyl ketones and lactones are formed. These components are almost absent in skim milk powder (low fat content). Dry powder obtained from spray drying has less cooked flavor than drum dried.

## 6.5 FACTORS AFFECTING KEEPING QUALITY OF MILK POWDER PRODUCTS

Keeping quality refers to the duration up to which edible qualities of food are retained. During the storage of dried milk powder and milk products powders, changes occurs in color, flavor and in solubility which are influenced by composition of milk, initial microbial load in the milk, quality of milk, hygiene maintained during the manufacturing process, manufacturing conditions, type of packaging, and storage conditions. Powders rich in fat and moisture content have more chances of getting spoiled and reducing the keeping quality. Moisture content should be maintained below 4%. Heat treatment should be effective to remove all the micro-organisms present in the milk especially heat resistant bacteria and sore formers. Hygiene is very important and high environment count leads to contamination in the product. Ignorance during cleaning of machinery, equipments, and vessels may also cause contamination of the product. During the manufacturing process, there are chances of growth of thermophiles bacteria around 45°C (when heated in counter flow heat exchanger) but could not, multiply over 50°C. If packaging is carried out in contaminated environment or packaging material is not able to prevent moisture absorption and high level of oxygen inside the package, then there will be favorable conditions for proliferation of micro-organisms. Therefore, these factors should be taken into consideration while preparing powder product.

## 6.6 CHANGES DURING STORAGE OF MILK POWDER PRODUCTS

Oxidation of fat, hydrolytic rancidity, changes in lactose and proteins are observed during the storage of milk powder products. Decomposition of fat can be accelerated due to its oxidation. Oxidation of fat can be influenced by the storage temperature, presence of copper and iron, acidity, and sunlight.[3] These can be prevented by addition of antioxidants, lowering storage temperature, preheating, and gas packaging. Due to lactose protein changes off flavor and insolubility increases. Off-color is basically developed due to milliard reaction (lactose and protein interaction) and due to use of high temperature during drying. High protein denaturation also reduces solubility of powder.

## KEYWORDS

- density
- drum drying
- spray drying
- steam

## REFERENCES

1. Augustin, M. A.; Clarke, P. T. Dry Milk Products. In *Dairy Processing and Quality Assurance;* Chandan, R. C., Kilara, A., Shah, N. P., Eds.; John Wiley and Sons: Chichester, West Sussex, 2008; pp 319–336.
2. Chavan, R. S.; Sehrawat, R.; Mishra, V.; Bhatt, S. Milk: Processing of Milk. In *The Encyclopedia of Food and Health;* Caballero, B., Finglas, P., Toldrá, Eds.; Academic Press: Oxford, 2016; pp 729–735.
3. Kessler, H. G. *Food Engineering and Dairy Technology;* Verlag A. Kessler: Freising, West Germany, 1981; p 231.
4. Schuck, P. Milk Powder: Types and Manufacture. In *Encyclopedia of Dairy Science,* 2nd ed.; Fuquay, J. W., Fox, P. F., McSweeney, H., Eds.; Elsevier: London, 2011; pp 108–117.
5. Sharma, A.; Jana, A. H.; Chavan, R. S. Functionality of Milk Powders and Milk-based Powders for End Use Applications—A Review. *Compr. Rev. Food Sci. Food Saf.* **2012,** *11* (5), 518–528.
6. Skanderby, M.; Westergaard, V.; Partridge, A.; Muir, D. D. Dried Milk Products. In *Dairy Powders and Concentrated Milk Products*; Tamime, A. Y., Ed.; Blackwell Publishers: Oxford, UK, 2009; pp 231–245.
7. Sukumar, De, Ed. *Outlines of Dairy Technology;* Oxford University Press: Oxford, 2012; p 231.
8. Tufail, A., Ed. *Dairy Plant Engineering and Management;* Kitab Mahal Agencies: Allahabad, 2003; p 112.

9. Verdurmen, R. E. M.; de Jong, P. Optimizing Product Quality and Process Control for Powdered Dairy Products. In *Dairy Processing;* Smit, G., Ed.; CRC Press: Boca Raton, FL, 2003; pp 334–364.

10. Verdurmen, R. E. M.; Verschueren, M.; Gunsing, M.; Straatsma, H.; Blei, S.; Sommerfeld, M. Simulation of Agglomeration in Spray Dryers: The EDECAD Project. *Lait.* **2005,** *85,* 343–351.

## CHAPTER 7

# OSMOTIC DEHYDRATION: PRINCIPLES AND APPLICATIONS IN THE DAIRY INDUSTRY

ADARSH M. KALLA* and DEVARAJU RAJANNA

*Department of Dairy Engineering, Dairy Science College, Karnataka Veterinary Animal and Fisheries Sciences University, Mahagoan Cross, Kalaburagi 585316, Karnataka, India*

*Corresponding author. E-mail: adarshkalla002@gmail.com*

## CONTENTS

## ABSTRACT

Osmotic dehydration is a process of water removal by immersion of water containing cellular solids in a concentrated aqueous solution of sugar/salt. The process is achieved by direct contact of fruits with a hypertonic solution. Water removal in this process is carried out without phase change. Osmotic dehydration is more than a preservation technique because of the superior quality of the finished product. Besides, it improves the color, flavor, and texture and is less energy intensive process compared to other drying techniques. The technique of osmotic dehydration is gaining popularity as a complementary processing step. Generally, osmotic dehydration is inherently slow and there is a need to find ways of increasing mass transfer, but due to its energy and quality related advantages it has gained recognition.

## 7.1 INTRODUCTION

Water is recognized as being one of the very important constituent of food. Its presence is very critical and affects food stability, microbial as well as chemical, and is responsible for the consumer perception of many organoleptic attributes, that is, juiciness, elasticity, tenderness, and texture. It is generally accepted that it is not the quantity of water in food but its thermodynamic state that is responsible for its influence on food stability and texture. The thermodynamic state of water in food is expressed by its activity, which is zero for absolutely dry material and one for pure water. In most of cases, lower the water activity, more stable is the food. The texture changes from juicy and elastic to brittle, and crunchy. Controlling the water within a product, by some method of drying or by chemically/ structurally binding (salting or sugaring), has long been used by humans for preservation. This not only controls microbial spoilage, but also chemical and physical stability of food material. Various drying methods have been developed but most of them are energy intensive, and hence the final product is expensive. The use of osmosis allows both way of reducing the water activity and is cost effective.

Osmotic dehydration is a process of water removal from plant tissues by immersion in a hypertonic (osmotic) solution. Water removal is based on the natural and non-destructive phenomenon of osmosis across cell membranes. The driving force for the diffusion of water from the tissue into the solution is provided by the higher osmotic pressure of the hypertonic solution. The permeability of plant tissue is low to sugars and high molecular weight compounds. Hence, the material is impregnated with the osmoactive substance in the surface layers only. Water on the other hand, is removed by osmosis and the cell sap is concentrated without a phase transition of the solvent. This makes the process favorable from the energetic point of view. The flux of water is much larger than the countercurrent flux of osmoactive substance. For this reason, the process is called osmotic dehydration or osmotic dewatering.

The diffusion of water is accompanied by the simultaneous counter diffusion of solutes from the osmotic solution into the tissue. Since the cell membrane responsible for osmotic transport is not perfectly selective, solutes present in the cells (organic acids, reducing sugars, minerals, flavors, and pigment compounds) can also be leached into the osmotic solution,[11,15] which affect the organoleptic and nutritional characteristics of the product. The rate of diffusion of water from any material made up of such tissues depends upon factors such as temperature and concentration of the osmotic solution, the size, and geometry of the material, the solution-to-material mass ratio, and to a certain level agitation of the solution.[42,54] Mass transfer during osmotic treatment occurs through semipermeable cell membranes of biological materials, which offers dominant resistance to the mass transfer. The state of the cell membrane can change from partial to total permeability by heat treatment and this can lead to significant changes in the tissue architecture. During osmotic removal of water from foods, the dehydration front moves from the food surface to the center. The osmotic stress results in cell disintegration and causes cell damage. It is attributed to the reduction in size caused due to water loss, resulting in the loss of contact between the outer cell membrane and the cell wall.[44]

The difference in the respective chemical potentials leads to out flow of water from the product and solute inflow into the product from osmotic solution, which mainly occurs in the first 2–3 h of immersion. After that the water content between food and osmotic solution gradually decreases, until eventually the system reaches a state of dynamic equilibrium of molecule transfer.[48]

The drying is a most widely used preservative method for many agricultural products. However, conventional drying methods are energy intensive, as they involve simultaneous heat and mass transfers. The exhaustion of non-renewable energy sources has led to expensive energy sources and ultimately increased process cost. Moreover, the use of high temperatures can result in the degradation and oxidation of some food nutrients. Thus, osmotic dehydration stands out as an efficient pre-treatment, improving the final product's quality, due to a reduction in nutrients degradation such as vitamins and minerals. It also shows other advantages such as energy saving (reduction in drying time) and the fact that the product is processed in a liquid phase, allowing for good heat and mass transfer coefficients.[42] Osmotic dehydration is used as a pre-treatment to many processes such as freezing, freeze drying, vacuum drying, or air drying as it improves the nutritional, sensorial, and functional properties of food without changing its integrity. It also increases the sugar to acid ratio and improves the texture and stability of pigments during dehydration as well as storage.[42] The osmotic process may provide the possibility of modifying the functional properties of food materials, improving the overall quality of the final product, and giving potential energy savings.[48] The other major application is to reduce the water activity of many food materials so that microbial growth will be inhibited.

Osmotic dehydration has received greater attention in recent years as an effective method for preservation of fruits and vegetables. In the hypertonic solution, osmatic substance in fruits releases water and natural soluble compounds. Being a simple process, it facilitates processing of tropical fruits and vegetables such as banana, sapota, pineapple, mango, and leafy vegetables etc., with retention of initial fruit and vegetables characteristics, namely, color, aroma, and nutritional compounds.[37]

## 7.2   MECHANISM OF OSMOTIC DEHYDRATION

In osmotic dehydration, the mass transfer occurs through the semipermeable cell membrane of the biological food material, which offers a dominant resistance to mass transfer. Diffusion of water takes place through the semipermeable membranes of fruit and vegetable tissues. The changes occurring in the state of the cell membranes were not considered in most of the earlier studies, although the diffusion coefficients for water have been generally considered to be constant throughout the process. Since cell membrane properties change due to osmotic stress, the models previously reported are not very appropriate. The state of the cell membrane can change from partial to total permeability, leading to significant changes in the rate of mass (water) transfer across it.

The difference between the mechanisms of osmotic water removal from homogeneous materials and from cellular biological materials[45] has been explained by Rastogi et al.[45] In homogeneous material, it is generally assumed that constant rate diffusion (with diffusion coefficient, D) occurs under the influence of a uniform moisture gradient. However, this does not appear to be true, especially after the initial stages of the process are over and the physical structure of the material starts to change.

In the mechanism for cellular biological materials,[45] it is proposed that the dehydration front (represented by $\Delta x$) moves toward the center of the material. This results in cell membrane disintegration in the dehydrated region and the water is transported across three different regions (each with its own characteristic properties): Diffusion of water from the core of the material to the dehydration front; diffusion of water across the front; and diffusion of water through the osmotically dehydrated layers into the surrounding medium.

At first, water diffuses from the outer layer of the sample to the osmotic medium, thereby increasing the osmotic pressure at the surface. As the osmotic pressure reaches a critical value, the cell membranes shrink and rupture. This results in a steep reduction in the proportion of intact cells, which is reflected in an increase in the cell permeabilization index (Zp). The condition of the cell or the degree of disintegration can be examined by Zp, that is, measured by electrophysical measurement based on electrical impedance analysis.[19] In other words, Zp is an integral parameter, which indicates the relative reduction in the proportion of intact cells. The diffusion coefficient of water released through these ruptured and shrunk cells into the osmotic solution at any point of time is denoted by $D_3$.

As osmotic dehydration proceeds, the dehydration front moves into the product. In this front, the cells are in the process of disintegration and hence, the rate of mass transfer increases sharply. At this juncture, a relatively large amount of water diffuses out with a diffusion coefficient $D_2$ ($D_2 \gg D_3$). As the cells in the core of material are intact, the diffusion coefficient of water from this core ($D_1$) is much lower than $D_2$ and $D_3$. The representative profiles for the cell disintegration index (Zp) and relative moisture content (M/Mo) values are also shown in a research article by Rastogi et al.[46]

## 7.3    MASS TRANSFER IN OSMOTIC DEHYDRATION

The plant cell generally consists of two main components: cell wall and protoplast. The cell wall forms the outer lining of the cell and covers the cell membrane. It is perforated and permeable to water and solute particles. The protoplast contains the protoplasm enclosed in a membrane, and it also constitutes other structural elements such as nucleus, vacuoles, plastids, and so on. Protoplasm is a colloidal solution of proteins and lipoproteins in water. The vacuole is suspended in protoplasm and is enclosed in a membrane called the tonoplast. It contains a solution of minerals, sugars, and other organic compounds in water.

Plant cells maintain a delicate balance of water, various dissolved salts, and sugars. It is important for plant cell to maintain the water pressure within the cells, keeping them rigid so they can support the plant. The water pressure within the plant cell is called as turgor pressure that is maintained by a process called osmosis. The fluid inside the plant cell is rich in solute concentration than the surrounding fluid. The water moves inside and tries to reach equilibrium, meanwhile at the same time the solute particles also try to diffuse out until they reach equilibrium. However, this does not happen as the cell membranes of plant cells are semi-permeable, it allows only water molecules to diffuse through it, since the solute particles are too large to pass through them. This causes to build up water pressure inside the plant cell, causing the cell membrane to exert pressure on the cell wall (In fact the content of vacuole increases, which has osmotic pressure that pushes protoplasm toward the cell wall). The protoplast is tightly pressed to the cell wall and the cell is in a turgor state. The difference between the osmotic pressure in the cell and in its surroundings is called the turgor pressure. Hence, the plant cells have to maintain their internal water pressure or turgor pressure, any imbalance in this pressure leads to collapse of the cell.

If the cell and the surroundings have, the same osmotic pressure then turgor pressure is zero and the system is in thermodynamic equilibrium. If osmotic pressure of the surroundings is lower than that of the cell, there is a transfer of water into the cell. The cell swells, but the rigid cell wall limits the extent of swelling. A cell placed in a hypertonic solution (osmotic pressure higher than that of the cell) will lose water. The dehydration of a protoplast causes decrease of its volume. This process is called plasmolysis. As the cell wall is permeable, the volume between the cell wall and cell membrane fills with the hypertonic solution.

Osmotic dehydration occurs on a piece of material and not on a single cell. Hence, it should be assumed that the piece exists in all kinds of plant tissue. From the process point of view, a plant material can be considered as a capillary-porous body that is divided internally in numerous repeating units. Some capillaries and pores are filled with a solution, whereas others are empty (i.e., contain air). Most capillaries and pores are open. Repeating units can exchange water between each other. The internal structure of a body is not a homogeneous one as far as transport of water is considered. Cell walls are built from microfibrils and intermicrofibrillar spaces. These spaces are large enough to allow water, ions, and small molecules to pass through them. As cell walls are interconnected in the tissue, a continuous matrix capable of transporting water and small molecules is formed. This continuum is called the apoplast.

## 7.4 FACTORS EFFECTING OSMOTIC DEHYDRATION

Several factors can affect the mass transfer of water and solute in osmotic dehydration such as: the temperature of osmoactive substance, osmoactive solution concentration, fruit-solution ratio, agitation level, processing time, and temperature of solution. Temperature is one of the most significant parameters in the osmotic dehydration kinetics. In general, the higher the process temperature, the greater the mass transfer rate, mainly due to the increase in cell permeability.[54] The increased dehydrating solution concentration will favor mass transfer; hence, solutions close to saturation can promote an increase in the water loss.[54] In most cases, this solution ends up being discarded after dehydration, and thus it is of interest to increase the fruit-solution ratio as much as possible or the osmotic solution needs to be re-concentrated by some means, either by evaporation or by adding fresh osmotic reagent. This makes osmotic dehydration more attractive in economic terms.

### 7.4.1 RAW MATERIAL CHARACTERISTICS

#### 7.4.1.1 QUALITY OF RAW MATERIAL

The different species, different varieties of same species, and maturity of fruits and vegetables mainly control water loss and solid gain in the osmosis process. Among different fruits, variability is mainly related to the natural tissue structure in terms of cell membrane structure, tissue compactness, initial insoluble and soluble solids content, intercellular spaces, and enzymatic activity of the fruit. These structural differences substantially affect diffusional mass exchange between the product and osmotic medium. Hartal[16] showed that under identical process conditions different potato varieties give substantially different weight reduction (water loss). Among different varieties of mango, Dashehari, and Totapuri at ripe stage were found suitable for osmotic dehydration.[53]

### 7.4.1.2   GEOMETRY OF THE MATERIAL

The geometry of sample pieces affects the behavior of the osmotic dehydration due to the variation of the surface area per unit volume (or mass) and diffusion length of water and solutes involved in mass transport.[24] In an investigation, it was found that only up to a certain total surface area/half thickness (A/L) ratio, higher specific surface area sample shape (such as rings) gave higher water loss and sugar gain value compared to lower surface area samples (such as slices and stick).[24] Exceeding this A/L limit, however, higher specific surface area samples (such as cubes) favored sugar gain at the expense of lower water loss resulting in lower weight reduction. The lowest water loss association with the highest A/L ratio was explained as a result of reduced water diffusion due to the high sugar uptake.

### 7.4.2   OSMOTIC PROCESS PARAMETERS

#### 7.4.2.1   TEMPERATURE OF THE OSMOACTIVE SOLUTION

The most important variable affecting the kinetics of mass transfer during osmotic dehydration is temperature. The increase in temperature enhances the loss of water but has little effect on the uptake of solid into the feed.[4]

After evaluation of the effect of temperature on the osmotic dehydration of star-fruit and kiwi, respectively, it was reported that some alterations can occur at higher temperatures such as swelling and plasticization of the cell membrane, which make the fruits more permeable to the passage of water and entrance of solids.[3,34] In addition, an increase in temperature results in a decrease in the viscosity of the osmotic solution reducing the external resistance to mass transfer.[9] The process carried out at 50°C for star-fruit resulted in a product with lower solid gain and greater water loss. This is probably due to the swelling and plasticization suffered by the cell membrane during osmotic dehydration at higher temperatures, which led to an increase in its permeability and a reduction in the viscosity of the osmotic solution.[5]

Many research studies on osmotic treatment have cited that a temperature around 50°C gave good results for osmotic dehydration of many vegetables and fruits due to the subsequent reasons: (1) this reasonable temperature has less deterioration of flavor, texture, and thermo sensible compounds of the materials, (2) the enzymatic browning and flavor deterioration of fruits start at temperature of 49°C, and (3) this temperature was also efficient to maintain the viscosity of the solution and adequate infusion time without changing the fruit quality.[1] It was reported that undesirable changes appeared on the blue berries at temperature of more than 50°C.[41,49]

## 7.4.2.2   CONCENTRATION OF SOLUTION

The increase in the osmotic solution concentration leads to an increase in water loss and solids gain of the fruit and vegetable cells, which could be explained by the greater concentration gradient between the fruit and the solution resulting in a greater dehydration driving force, which was observed by Teles[52] in the osmotic dehydration of melons at 65°C. Campos et al.[5] observed that processing of star-fruit at a solution concentration of 50% resulted in a product with lower solids gain and greater water loss. Increase in osmotic solution concentration resulted in corresponding increases in water loss to equilibrium level and drying rate.[8,21] Therefore, increased osmotic solution concentrations leads to increased weight reductions. This was attributed to the water activity of the osmotic solution which decreases with the increase in solute concentration in the osmotic solution.[30,55]

Studies by Saurel[47] showed a dense solute–barrier layer formed at the surface of the food material when the osmotic solution concentration increased. This enhanced the dewatering effect and reduced the loss of nutrients during the process. A similar solute–barrier is also formed in the case of osmotic solutions with higher molecular weight solutes even at low concentration. The syrup strength in the range of 60–70° Brix has been found to be optimum.[6] It was also reported that higher the concentration, faster the rate of osmosis. Torregiani[54] suggested that it was usually not worthwhile using higher concentration for osmosis process for more than 50% of weight reduction because of decrease in osmotic rate with time.

## 7.4.2.3   TYPE OF OSMOACTIVE AGENT

The selection of proper osmoactive agent is of great importance in osmotic dehydration process. Osmoactive substances used in food must comply with special requirements. They have to be edible with accepted taste and flavor, nontoxic, inert to food components, and if possible highly osmoactive. Solutions of sugars are mostly used to dehydrate fruits and glycerol, starch syrup, and sodium chloride are used for vegetables.[28,27] Sucrose is the most frequently used substance[18,38,40] and can be substituted in part by lactose.[17] The control of pH of sucrose solution is very important in osmotic dehydration as it affects the course of osmotic dehydration in various fruits and vegetables; and the control of pH is recommended for apple and carrot.[26] Addition of ascorbic acid to sugar solution has been practiced to minimize browning of fruit pieces during osmotic process. Mixtures of osmoactive substances are also used in hypertonic solution for osmotic dehydration. Sugar and salt solutions have proved to be the best choices based on effectiveness, convenience, and flavor. Experiments were conducted on dehydration of apples in a solution containing 42% fructose, 52% sucrose, 3% maltose, 3% polysaccharides, and 0.5% sodium chloride in dry matter.[25] Use of sucrose and starch syrup has been reported to be used in a ratio of 1:1.[29] The studies were conducted using solutions containing sucrose and fructose in varying proportions. Water loss was similar for all solutions under study, but the penetration of the osmoactive substance was different.

The comparative study between various osmotic solutions at constant solid concentration reported that mixed sucrose and salt solutions gave a greater decrease in product water activity compared to pure sucrose solutions, although water transport rates were similar,[22] due to extensive salt uptake. Further studies by the same workers on spatial distribution analysis revealed large differences between osmosis distribution curves for the dehydration taking place in sucrose or salt solutions.[23] Their analysis showed that sucrose accumulated in the thin sub-surface layer resulting in surface tissue compacting (an extra mass transport barrier) and salt was found to penetrate the osmosed tissue to a much greater depth. The presence of salt in the osmotic solution can hinder the formation of the compact surface layer, allowing higher rates of water loss and solid gain. Finally, increasing salt concentration leads to a lower water activity solution with increased driving (osmotic) force. Peach fruits dehydrated in solutions of glucose and fructose was especially suitable for pasteurization. Solution of sucrose and glucose yielded high drying rate of apple slices.[18] Mixture of crystalline sucrose and glucose lowered water activity of guava to 0.77 whereas sucrose alone yielded a water activity of 0.80.[2]

## 7.4.2.4   AGITATION OF OSMOTIC SOLUTION

The osmotic dehydration is enhanced by agitation or circulation of the osmotic solution around the sample. Agitation ensures a continuous contact of the sample surface with concentrated osmotic solution, securing a large gradient at the product/solution interface.[7,17] The agitation has a tremendous impact on weight loss; water removal is affected by high viscosity of the solution causing resistance to mass transfer, floating of food pieces in solution eventually hindering the contact between food material and the osmotic solution.[32,35] The agitation can increase the turbulence which increases the mass transfer.

In some studies on osmotic dehydration, it was observed that agitation favors water loss, especially at lower temperatures (< 30°C), where viscosity is high and during the early stages of osmosis.[43] The extent of water loss is increased with agitation and reached a certain plateau. On the other hand, the rate of solid gain is decreased with agitation. For short process periods, agitation has no effect on the solid gain. For longer process period, solid gain decreased drastically with agitation. The authors concluded that agitation has no direct impact on solid gain throughout the entire osmotic process, since external transfer of the osmotic solute is not limiting.

The studies were conducted to compare the effects of agitation and non-agitation treatments.[32] The agitated samples exhibited greater weight reduction, accordingly water loss, than non-agitated samples. The agitation or stirring process can promote the turbulent flow, resulting in the increment of liquid diffusion during osmotic dehydration. Turbulent flow can enhance the hydrodynamic flow mechanism during osmotic dehydration. Therefore, the agitation or stirring process could be a good alternative way to enhance mass transfer, leading to the reduction of the contact time to achieve determined moisture content in the food materials.[35]

## 7.4.2.5 FRUIT TO OSMOTIC SOLUTION RATIO

The process carried out with a smaller amount of solution has lower water loss. This could be attributed to the fact that the osmotic solution was diluted by the water removed from the product resulting in a smaller concentration gradient and, consequently, a reduced water loss.[5] Greater water loss and solids gain were reported when the fruit to syrup ratio varied from 1:2 to 1:4 after 1.5 h of osmotic dehydration of bananas under vacuum.[51] However, when the process was carried out at atmospheric pressure, the increase in the fruit to syrup ratio did not result in significant differences in the water loss or solids gain. An increase of osmotic solution to sample mass ratio resulted in an increase in both the solid gain and water loss in osmotic dehydration.[14,39] To avoid significant dilution of the medium and subsequent decrease of the (osmotic) driving force during the process, a large ratio has been used by most investigators.

## 7.5  EQUIPMENTS FOR OSMOTIC DEHYDRATION PROCESSING

Osmotic dehydration is a non-thermal dehydration process, more efficient than the conventional drying operation. It is used to produce intermediate–moisture foods, as pre-drying step to remove 50% moisture from the foods. Equipment for osmotic dehydration is in development stage before large-scale commercial applications. It may be similar to extraction or leaching equipment, used in chemical engineering. While designing equipment for osmotic dehydration, the product characteristics and processing parameters have to be taken into account. The equipment needed must assure control of processing parameters, efficiency, and economics as well. Choice of the equipment is based on the following criteria:[36]

- Type of processing: periodic or continuous.
- Resistance of food to mechanical damage.
- Shape of food: whole or cut into pieces.

- Susceptibility of food to oxidation in contact with air.
- Relative movement of phases: solid and liquid.
- Possibility to control processing parameters.
- Investment and running cost.
- The processes of osmotic dehydration can be categorized as follows:[31]
- Those in which food is immersed in the osmotic solution.
- Those in which solution is introduced onto the food.
- Those in which osmotic substance in solid state is contacted with food.
- Those in which reduced pressure is used to facilitate mass transfer.

## 7.6   FOOD IMMERSED IN SOLUTION

The simplest way to contact food with osmotic solution is to immerse a basket with food into solution. The movement of solution is slow due to natural convection. The mass transfer is also slow and most of the processing parameters are also not controlled. The method can be used for soft fruits. Osmotic dewatering can be facilitated by decreasing mass transfer resistance. This can be done either by circulation of solution or by slow movement of food. Circulation of solution is done by installation of circulation pump in a vessel in which basket with food is immersed. Movement of food in the solution is done by vibration or by a conveyor. The latter method is used in Poland in the processing of apple slices. Combination of solution circulation and movement of food particles is combined in equipment such as vibrating plate mixer and percolated bed with slow displacement of food.

In vibrated plate mixer,[36] osmotic solution is circulating in two loops, one is a feed loop by which food is fed into a mixer, and the second loop maintains constant temperature of the solution. Food moves from bottom of the mixer to its top through a series of perforated vibrating horizontal plates mounted on a vertical axis.

In a percolated bed, food is delivered at the bottom of the tank by a hydraulic feed and forms a porous bed. The bed moves slowly to the top of the tank and is extracted by a bucket conveyor. Solution is fed at the top of the tank and is circulated through the feed leg. In this equipment, a countercurrent movement of food and solution occurs. A concurrent movement of food and solution as a percolated bed was also designed for osmotic process.[31]

Movement of solution and particles can be done by mechanical mixing. Mixing device can be installed vertically or horizontally. In the first design, a worm screw is placed coaxially inside a vertical cylindrical tank. The screw moves particles of food from top to the bottom of the tank. Then the pieces rise toward the surface under the buoyancy force. In the second technical solution or design, the screw is mounted horizontally. The food pieces together with the solution are moved along the cylinder axle. Pieces of food are carried in rotation toward the end of the cylinder where a deflector catches the pieces and directs them to the outlet.[36]

Designs of equipment with mechanical motion of food pieces exert some force on processed material. Hence, some disintegration and deformation of food can take place and increased pulp content in the solution can be observed.

## 7.7   SOLUTION SPRAYED ONTO THE FOOD

Reduction of solution and food ratio can be done by application of thin layer of hypertonic solution to food pieces. It is done by placing food pieces on perforated conveyor and spraying concentrated solution on processed material. The design is well suited to continuous processing but requires food pieces to be spread on the conveyor in a single layer. Hence, a large area of the conveyor is needed to process any given quantity of food. This technical solution of osmotic watering has been proposed.[10,20]

## 7.8   SOLID OSMOTIC SUBSTANCE CONTACTED WITH FOOD

The lowest solution and food ratio is obtained when solid osmotic substance is contacted with food. Crystals of sugar or mixture of sugar and salt are mixed with food pieces in appropriate proportion and tumbled in slowly rotating cylindrical tank. The amount of osmotic substance used should be such that water removed from food pieces forms no solution in the tank. Wet but solid osmotic substance is separated from food on vibrating screen. However, some crystals stick to the food surface and can create problems in packaging or further processing of osmatic material.

## 7.9   EQUIPMENT WORKING UNDER REDUCED PRESSURE

The vacuum impregnation of the porous product consists of exchanging the internal gas or liquid occluded in open pores for an external liquid phase, due to the action of hydrodynamic mechanisms promoted by pressure changes.[12,13] The effect of continuous and pulsed vacuum was determined for the osmotic dehydration rate of apricots, strawberries, and pineapples.[50] Experiments in 65° Brix sucrose solutions at 30–50°C temperatures for 15–240 min under atmospheric pressure, vacuum (100 mbar) and pulsed vacuum (100 mbar for 5 min) were conducted. It was found that the vacuum treatments have a significant effect on water transfer rate, but had no effect on sugar uptake between vacuum and normal pressure treatments during osmotic dehydration. The pulsed vacuum treatment was found to be highly economical and the influence of vacuum treatments on water loss was more effective on pineapple because of their higher porosity. Experiments were conducted on pulsed vacuum osmotic dehydration of strawberries at 50 mbar for 5 min, after which it was continued under atmospheric pressure.[33] It was reported that the pulsed vacuum osmotic dehydration increases the mass transfer. Static or pulsed-vacuum processing of immersed fruit or vegetable products facilitates osmotic dewatering. Equipment used in this process can be of any type previously presented but requires hermetic design.

## 7.10   CONCLUSIONS

During last few decades, research studies on osmotic dehydration have given a fruitful result for food industry. The osmotic dehydration are mostly used to produce intermediate moisture foods and semidried fruits, which are used as ingredients in many complex foods like icecream, cereals, confectionery, and dairy products. Osmotic dehydration can be considered as the most eligible energy saving method for water removal. It is a non-thermal process and hence, can be used for processing the heat sensitive products and secure the quality of food. The osmotic dehydration reduces the moisture content of foods before they are subjected to further processing steps such as drying, freezing, frying, etc. The osmotic dehydration efficiency can be increased by controlling the factors (namely, concentration, temperature, type of osmoactive substance, agitation of the solution, etc.) affecting the mass transfer during the osmotic dehydration process. Although, various equipments have been designed for processing of agricultural product by osmotic dehydration; however, continuous and automated processing equipments are required for large scale processing by osmotic dehydration of food materials.

## KEYWORDS

- **cell disintegration index**
- **diffusion coefficient**
- **hypertonic solution**
- **osmotic dehydration**
- **thermodynamic equilibrium**
- **turgor pressure**
- **vibrating mixer**

## REFERNCES

1. Akbarian, M.; Ghasemkhani, N.; Moayedi, F. Osmotic Dehydration of Fruits in Food Industrial: A Review. *Int. J. Biosci.* **2014,** *4* (1), 42–57.
2. Ayub, M.; Khan, R.; Wahab, S. Effect of Crystalline Sweeteners on the Water Activity and Shelf Stability of Osmotically Dehydrated Guava. *Sarhad J. Agric.* **1995,** *11,* 755.
3. Bauchot, A. D.; Hallett, I. C.; Redgwall, R. J.; Lallu, N. Cell Wall Properties of Kiwi-fruit Affected by Low Temperature Break Down. *Postharvest Biol. Technol.* **1999,** *16* (3), 245–255.
4. Beristain, C. L.; Azuara, E.; Cortes, R.; Garcia, H. S. Mass Transfer during Osmotic Dehydration of Pineapple Rings. *Int. J. Food Sci. Technol.* **1999,** *25,* 576–582.
5. Campos, M. D. C.; Sato, K. C. A.; Tonon, V. R.; Hubinger, D. M. Effect of Process Variables on the Osmotic Dehydration of Star-fruit Slices. *Cienc. Tecnol. Aliment Campinas.* **2012,** *32* (2), 357–365.
6. Chaudhary, A. P.; Kumbhar, B. K.; Singh, B. N. N.; Narain, M. Osmotic Dehydration of Fruits and Vegetables. *Indian Food Industry.* **1993,** *12,* 20–27.
7. Contreras, J. E.; Smyrl, T. G. An Evaluation of Osmotic Concentration of Apple Rings Using Corn Solids Solutions. *Can. Inst. Food Sci. Technol. J.* **1981,** *14,* 310–314.
8. Conway, J.; Castaigne, F.; Picard, G.; Vovan, X. Mass Transfer Considerations in the Osmotic Dehydration of Apples. *Can. Inst. Food Sci. Technol. J.* **1983,** *16,* 25–29.
9. Cussler, E. L. *Diffusion Mass Transfer in Fluid Systems;* University Press: Cambridge, NY, 1997; p 580.

10. Dalla-Rosa, M.; Bressa, F.; Mastrocola, D.; Pittia, P. Use of Osmotic Treatments to Improve the Quality of High Moisture-minimally Processed Fruits. In *Osmotic Dehydration of Fruits and Vegetables;* Lenart, A., Lewicki, P. P., Eds.; Warsaw Agricultural University Press: Warsaw, Poland, 1995; p 69.

11. Dixon, G. M.; Jen, J. J. Changes of Sugar and Acid in Osmovac Dried Apple Slices. *J. Food Sci.* **1997,** *42,* 1126–1131.

12. Fito, P.; Pastor, R. On Some Non-diffusional Mechanism Occurring during Vacuum Osmotic Dehydration. *J. Food Eng.* **1994,** *21,* 513–519.

13. Fito, P. Modeling of Vacuum Osmotic Dehydration of Food. *J. Food Eng.* **1994,** *22,* 313–328.

14. Flink, J. M. Dehydrated Carrot Slices: Influence of Osmotic Concentration on Drying Behaviour and Product Quality. In *Food Process Engineering;* Linko, P., Malki, Y., Olkku, J., Larinkari, J., Fito, P., Eds.; Applied Science Publishers: London, 1979; pp 412–418.

15. Giangiacomo, R.; Torreggiani, D.; Abbo, E. Osmotic Dehydration of Fruit. Part I: Sugar Exchange between Fruit and Extracting Syrup. *J. Food Process. Pres.* **1987,** *11,* 183–195.

16. Hartal, D. Osmotic Dehydration with Sodium Chloride and Other Agents. Ph.D. Thesis, University of Illinois, Urbana, IL, 1967.

17. Hawkes, J.; Flink, M. J. Osmotic Concentration of Fruit Slices Prior to Freeze Dehydration. *J. Food Process. Pres.* **1978,** *2,* 265–284.

18. Kaymak-Ertekin, F.; Sultanoglu, M. Modelling of Mass Transfer during Osmotic Dehydration of Apples. *J. Food Eng.* **2000,** *45,* 243.

19. Knorr, D.; Angersbach, A. Impact of High Electric Field Pulses on Plant Membrane Permeabilization. *Trends Food Sci. Technol.* **1998,** *9,* 185–191.

20. Le-Maguer, M. In *Osmotic Dehydration: Review and Future Directions,* Proceedings of the Symposium in Food Preservation Processes, CERIA, Brussels, Belgium, 1988; Vol. 1, p 283.

21. Lenart, A. Mathematical Modelling of Osmotic Dehydration of Apple and Carrot. *Polish J. Food Nutr. Sci.* **1992,** *1,* 1–33.

22. Lenart, A.; Flink, J. M. Osmotic Concentration of Potato. I. Criteria for the End-point of Osmosis Process. *J. Food Technol.* **1984,** *19,* 45–60.

23. Lenart, A.; Flink, J. M. Osmotic Concentration of Potato. II. Spatial Distribution of the Osmotic Effect. *J. Food Technol.* **1984,** *19,* 65–89.

24. Lerici, C. R.; Pinnavia, G.; Rosa, M. D.; Bartolucci, L. Osmotic Dehydration of Fruit: Influence of Osmotic Agents on Drying Behavior and Product Quality Process. *J. Food Technol.* **1985,** *7,* 147–155.

25. Lerici, C. R.; Pepe, N.; Pinnavaia, G. The Fruit Dehydration by Direct Osmosis. I. Results of Experiments Carried Out in the Laboratory (La Disidratazione Della Frutta Mediante Osmosi Diretta. I. Resultati di Esperienze Effettuate in Laboratorio). *Ind. Conserve.* **1977,** *52,* 1.

26. Lewicki, P. P.; Lenart, A. The Effect of pH on the Course of Osmotic Dehydration. In *Osmotic Dehydration of Fruits and Vegetables;* Lewicki, P. P., Mazur, E., Eds.; Warsaw Agricultural University Press: Warsaw, Poland, 1995; p 20.

27. Lewicki, P. P.; Lenart, A.; Pakuła, W. Influence of Artificial Semi-permeable Membranes on the Process of Osmotic Dehydration of Apples. *Ann. Warsaw Agric. Univ. Food Technol. Nutr.* **1984,** *16,* 17.

28. Lewicki, P. P.; Lenart, A.; Turska, D. Diffusive Mass Transfer in Potato Tissue during Osmotic Dehydration. *Ann. Warsaw Agric. Univ. Food Technol. Nutr.* **1984,** *16,* 25.

29. Maltini, E.; Torreggiani, D.; Bertolo, G.; Stecchini, M. In *Recent Developments in the Production of Shelf-stable Fruit by Osmosis,* Proceedings of 6th International Congress Food Science and Technology, Dublin, Ireland, 1983; p 177.

30. Marcotte, M.; Toupin, C. J.; Le Maguer, M. Mass Transfer in Cellular Tissues. Part I: The Mathematical Model. *J. Food Eng.* **1991,** *13,* 199–220.

31. Marouze, A.; Giroux, F.; Collignan, A.; Riviero, M. Equipment Design for Osmotic Treatments. *J. Food Eng.* **2001,** *49,* 207.

32. Moreira, R.; Chenlo, F.; Torres, M. D.; Vazquez, G. Effect of Stirring in the Osmotic Dehydration of Chestnut Using Glycerol Solutions. *LWT-Food Sci. Technol.* **2007,** *40,* 1507–1514.

33. Moreno, J.; Chiralt, A.; Escriche, I.; Serra, J. A. Effect of Blanching/Osmotic Dehydration Combined Methods on Quality and Stability of Minimally Processed Strawberries. *Food Res. Int.* **2000,** *33,* 609–616.

34. Perez-tello, G. O.; Silva-Espinoza, B. A.; Vargas-Arispuro, I. Effect of Temperature on Enzymatic and Physiological Factors Related to Chilling Injury in Carambola Fruit (*Averrhoa carambola* L.). *Biochem. Biophys. Res. Commun.* **2001,** *287* (4), 846–851.

35. Phisut, N. Factors Affecting Mass Transfer during Osmotic Dehydration of Fruits, *Int. Food Res. J.* **2012,** *19* (1), 7–18.

36. Piotr, P.; Lewicki, C.; Lenart, A. Osmotic Dehydration of Fruits and Vegetables. In *Handbook of Industrial Drying,* 3rd ed.; Mujumdar, S. A., Ed.; CRC Press: Oxon, UK, 2006; pp 687–709.

37. Pokharkar, S. M.; Prasad, S. Mass Transfer during Osmotic Dehydration of Banana Slices. *J. Food Sci. Technol.* **1998,** *35* (4), 336–338.

38. Ponting, J. D. Osmotic Dehydration of Fruits—Recent Modifications and Applications. *Process. Biochem.* **1973,** *8,* 18.

39. Ponting, J. D.; Watters, G. G.; Forrey, R. R.; Jackson, R.; Stanley, W. L. Osmotic Dehydration of Fruits. *Food Technol.* **1966,** *20,* 125–128.

40. Quintero-Ramos, A.; de la Vega, C.; Hernandez, E.; Anzaldua-Morales, A. (1993). Effect of the Conditions of Osmotic Treatment on the Quality of Dried Apple Dices. *AIChE Symp. Ser.* **1993,** *89* (297), 108.

41. Rafiq, K. M. Osmotic Dehydration Technique for Fruits Preservation—A Review. *Pakistan J. Food Sci.* **2012,** *22* (2), 71–85.

42. Raoult-Wack, A. L. Recent Advances in the Osmotic Dehydration of Foods. *Trends Food Sci. Technol.* **1994,** *5* (8), 255–260.

43. Raoult-Wack, A. L.; Lafont, F.; Rios, G.; Guilbert, S. Osmotic Dehydration: Study of the Mass Transfer in Terms of Engineering Properties. In *Drying;* Mujumdar, A. S., Roques, M., Eds.; Hemisphere Publishing Corporation: New York, NY, 1988; Vol. 89, pp 487–495.

44. Rastogi, N. K.; Angersbach, A.; Knorr, D. Synergistic Effect of High Hydrostatic Pressure Pretreatment and Osmotic Stress on Mass Transfer during Osmotic Dehydration. *J. Food Eng.* **2000,** *45* (1), 25–31.

45. Rastogi, N. K.; Angersbach, A.; Knorr, D. Evaluation of Mass Transfer Mechanisms during Osmotic Treatment of Plant Materials. *J. Food Sci.* **2000,** *65* (6), 1016–1021.

46. Rastogi, N. K.; Raghavarao, K. S. M. S.; Niranjan, K.; Knorr, D. Recent Developments in Osmotic Dehydration: Methods to Enhance Mass Transfer. *Trends Food Sci. Technol.* **2002,** *13* (2), 58–69.

47. Saurel, R.; Raoult-Wack, A.; Rios, G.; Guilbert, S. Mass Transfer Phenomena during Osmotic Dehydration of Apple. I: Fresh Plant Tissue. *Int. J. Food Sci. Technol.* **1994,** *29,* 531–537.

48. Shi, J.; LeMaguer, M. Osmotic Dehydration of Foods: Mass Transfer and Modeling Aspects. *Food Rev. Int.* **2002,** *18* (4), 305–335.

49. Shi, J.; Xue, J. S. Application and Development of Osmotic Dehydration Technology in Food Processing. In *Advances in Food Dehydration*; CRC Press: Boca Raton, FL, 2009.

50. Shi, X. Q.; Fito, P.; Chiralt, A. Influence of Vacuum Treatment on Mass Transfer during Osmotic Dehydration of Fruits. *Food Res. Int.* **1995,** *28,* 445–454.

51. Sousa, P. H. M.; Maia, G. A.; Souza, M. D. S. M.; Fifueiredo, R. W. D.; Nassu, R. T.; Souza-Neto, M. A. D. Influence of Concentration and Proportion Fruit: Syrup in the Osmotic Dehydration of Processed Bananas. *Sci. Technol. Food.* **2003,** *23,* 126–130.

52. Teles, U. M. Optimization of Osmotic Dehydration of Melons Followed by Air-drying. *Int. J. Food Sci. Technol.* **2006,** *41* (6), 674–680.

53. Tiwari, R. B.; Jalali, S. In *Studies on Osmatic Dehydration of Different Varieties of Mango,* Proceeding of First Indian Horticulture Congress, New Delhi, 2004.

54. Torregiani, D. Osmotic Dehydration in Fruit and Vegetable Processing. *Food Res. Int.* **1993,** *26* (1), 59–68.

55. Tortoe, C. A Review of Osmo-dehydration for Food Industry. *Afr. J. Food Sci.* **2010,** *4* (6), 303–324.

# PART III

# Energy Conservation: Opportunities in the Dairy Industry

# CHAPTER 8

# FOOD SAFETY AND MANAGEMENT SYSTEMS (FSMS): APPLICATIONS IN THE DAIRY INDUSTRY

ANGADI VISWANATHA[1*], C. RAM KUMAR[2], JAYEETA MITRA[3], and S. S. ROOPA[4]

[1]Department of Food Microbiology, College of Agriculture, Hassan, University of Agricultural Sciences, Bangalore 573225, India
[2]Department of Food Science and Technology, College of Agriculture, Hassan, University of Agricultural Sciences, Bangalore 573225, India
[3]Department of Agricultural and Food Engineering, Indian Institute of Technology, Kharagpur 721302, India
[4]Britannia Industries Ltd., Plot # 23, Bidadi Industrial Area, Bidadi Hobli, Ramanagara 562109, Karnataka, India
*Corresponding author. E-mail: angadigm@gmail.com

## CONTENTS

## ABSTRACT

This chapter reviews different aspects of food safety management used in a dairy processing unit. In the first section, the need for a food safety management system (FSMS) is discussed and followed by discussion of risk analysis and different critical control points used in various dairy products along with the benefits of FSMS.

## 8.1   INTRODUCTION

The dairy processing unit has conventionally ensured the safety and quality of dairy products through inspections and test methods at each step of processing under the label of quality control and assurance. The development of the dairy processing sector and the outburst of food poisoning issues made the researchers think on the improvement in current safety and quality assurance steps. However, safety of the food products remains a high priority for producers, veterinarians, consumers, and regulatory agencies.

India has been tops among the milk producing countries in the world since 1998, and it has largest bovine population in the world. In 1950–1951, milk in India was 17 million tons, which increased to 146.3 million tons in 2014–2015. India has the lion's share (18.54%) of world milk production (789 million tons in 2014). The per capita availability of milk in India is 322 g/day whereas the world average is 293.7 g/day.[3] Due to rapidly growing domestic demand and fluctuations in prices in international markets, India's share in global milk exports was 0.68% in 2013. Beside these reasons, it is also required to maintain high quality standards with consistency to export milk and milk products to other countries. Hence, the Food Safety Management System (FSMS) plays important role in achieving this objective.

While on one side, it is heartening to see the progress made in hazard analysis and critical control points (HACCP) implementation by certain industries, for example, dairy and marine, yet the situation in the overall food industry is still very dismissal. The implementation is taking place primarily in the organized sectors, which have a very small share in the overall industry, for example, the organized dairy sector just constitutes 15–16% of the entire dairy industry.

In milk processing plant, the raw milk is accepted based on its organoleptic and physicochemical properties. It is also necessary to test its bacteriological quality, somatic cell count, and antibiotic residues in raw milk by rapid methods to ensure superior quality fluid milk and its products.

The quality systems should be aimed at prevention of the defects and hazards in milk, rather than just their detection. In modern days the consumers, processors, and regulatory bodies show more interest on safe, wholesome, readily available milk products by emphasizing on better farm management, clean milk production, hygienic processing, storage, and distribution.[12]

Further, while India is the third largest food producer and second largest fruit and vegetable producer in the world, the processing amounts to a meager 2% (approx.) of the produce. In the Indian context, where the FSMS is in the developing stage, it requires commitment by the management, and thorough knowledge on the system, in order to implement FSMS in a phased manner across food sector, vendors, and particularly the unorganized sector.

This chapter discusses applications of FSMS in dairy industry.

## 8.2  IMPORTANCE OF PLANT SANITATION AND PERSONAL HYGIENE

The term "sanitation" includes all precautionary measures, required in the manufacturing, processing, storage, and distribution in order to ensure an unobjectionable, sound, palatable, and safe product, which is fit for human consumption.[5]

### 8.2.1  PLANT SANITATION

The plant sanitation is a critical area of FSMS to be of high importance in the dairy processing plant. Since milk is a perishable food and rich in nutrients, it gets spoiled easily with poor plant hygiene. Also unhygienic conditions of food contact surfaces of equipments lead to harboring deadly pathogens like *Listeria monocytogenes, Escherichia coli, Salmonella, Yersinia enterocolitica,* and *Campylobacter.* It may lead to products recall and economic losses and shall have negative impact on dairy business.

The equipments used for processing, storage, packaging the milk, and milk products should be cleaned and sanitized satisfactorily as per sanitary program. The soil found in dairy plant is rich in proteins, fats, lactose, and minerals.[5] The effective sanitary practices can remove soil, microbes through optimal combination of detergents, mechanical energy, water, and sanitizers. The clean-in-place (CIP) systems help in cleaning pipeline interiors, filters, storage tanks, plate heat exchangers, and cream separators effectively without dismantling. Depending on the load of soiling and usage of equipments, the cleaning solutions can be formulated and used either for single use or reuse in multi-cycles. The normal CIP cycle starts from pre-rinse, detergent cycle (alkali and acid) rinse, disinfection cycle, and final rinse with potable water.

The effective sanitation program should include sanitary standard operating procedures (SSOPs), which are written instructions for the employees. It should include:

- Name of the equipments and areas to be cleaned and sanitized.
- Instructions on cleaning and sanitizing procedure or method, effectively.
- Frequency of cleaning and responsibility.
- Check for effective cleaning by quality personnel.
- Records to be updated.
- Compliance with statutory regulations.

The plant equipments should be evaluated for presence of biofilms before designing sanitation programs. The biofilms act as source for spoilage organisms as well as pathogens leading to food poisoning and spoilage of finished product. In the in vitro and in situ studies, the sanitizer Iodophor at 10 ppm concentration, with contact time for 20 min, applied for a week, showed >3 log reduction in all segments of pasteurization lines.[18] The other sanitizers that can be used are steam, hot water (85°C), ozone, chemicals spraying, or fogging for closed containers.

## 8.2.2 PERSONAL HYGIENE

The personal cleanliness and behavior should be given urgent care in processing of dairy products. It is well established that many food-borne illness are caused by infected food handlers, who directly come in contact with, raw materials, packing materials, and finished goods during processing. The common microbial pathogens transmitted by infected food handlers are Norovirus, Hepatitis A, *Enterohemorrhagic E. coli, Shigella* spp., *Salmonella typhi*, and *Staphylococcus aureus*. Hence, it is necessary that workforce should be well trained in fundamental aspects of personal hygiene. This contamination spreads very easily as they are contagious in nature. The dairy products can be contaminated with above said microbial pathogens risking, vulnerable consumers like, infants, small children, immune-compromised people, pregnant women, and aged persons.[21]

Any person suffering from or a carrier of disease or pathogen should be restricted from entry into the processing area. He/she should report immediately about illness, symptoms of diseases, to supervisor or management. It helps supervisor to decide the work area allotment to such person/s or sanction leave as quarantine measures. Also, regular medical examination of food handlers to detect any communicable diseases, vision, hearing ability, and general health check-up should be a part of FSMS. The persons with wounds, cuts or open skin lesions should be refrained from processing area as the cuts and wounds may be source of pathogens. They can be covered with proper bandage to prevent any cross-contamination.

The food handlers should maintain good hand hygiene and sanitation. It is necessary to wash hands before touching the products, after using rest room, after handling waste, raw materials, or any chemical. He/she should exhibit good personal behavior and refrain from sneezing, smoking, coughing over materials, or products, chewing tobacco, nose blowing, and spitting inside the processing area. The work force should wear appropriate personal protective equipments, like aprons, headgear, gloves, and face masks during handling or packing the dairy products. The antiseptic, iso-propyl alcohol (70%) should be used after drying washed hands, so that it reduces/kills residual microbes left after hand washing.

The food handlers should not wear jewelry, watch, bindi, pins, buttons, and false nails should be discouraged from using strong deodorants and perfumes. The visitors to dairy plant should be guided properly about personal hygiene codes, should be given protective clothes, hair nets, shoe covers, and masks before entering the processing area. There should be approved jewelry policy, and visitor policy displayed at the entrance of processing area to create awareness about personal hygiene. Swab tests need to be conducted at regular intervals to assess the hand hygiene of food handlers, and awareness need to be created by sharing the results of swab test, also encouraging them with suitable rewards for adherence to hygienic practices.

### 8.2.3 BENEFITS AND APPLICATION OF HACCP

HACCP is a system, which identifies, evaluates, and controls the hazards that are significant for food safety. It is a tool, which guides the food handler to consider all safety aspects and implement them effectively in order to assure safety of the food. As such, HACCP system does not make food safe, but its effective application and implementation makes the difference. It should be used as a tool to market the food product; rather it should result in reducing the food-borne illness and rejections in the international trade, thereby gaining confidence of consumers.[5] The HACCP have many benefits, such as:

- Reduction in food-borne illness incidences.
- Increased confidence in providing safe food to consumers.
- Promotes international trade with increased confidence in food safety and enhanced access to international market.
- Helps in meeting legal requirements and improves public health.
- Effective use of resources, reduction in wastage as well as recall of the product.

Its food safety system, based on preventive measures and not like traditional approach to food control by finished product inspection and testing. It is a farm to fork approach embracing, the farmers/milk producers, processors, distributors, and retail agents. The successful implementation of HACCP is facilitated by commitment from top management. Each and every person involved in the application of system should be properly trained, made to understand their role, to fulfill their responsibilities effectively. It is applied in order to meet requirements of safe and wholesome dairy products production. It requires commitment from workforce and management.[13]

## 8.2.4   THE LOGICAL SEQUENCE OF APPLICATION OF HACCP

On September 1, 2005, International Organization for Standardization (ISO) published a standard, ISO 22000:2005a: FSMS Requirements, for organization in the food chain. It provides the framework of international requirements for global approach toward food safety. It emphasizes on implementation of FSMS by all types of organizations like feed producers, milk producers, transport, and storage operators, vendors of ingredients, packaging materials, and equipment suppliers.[4] It has adapted process management principles, based on Deming Cycle (Plan-Do-Check-Act), which may be broken down into various sub processes.[7] It provides an auditable standard, which can be used for internal audits, external audits, by public health authorities. It can be used as a tool to assess organization confirming its food safety policy, by planning, implementing, operating, and updating the existing FSMS.[15] The flow chart for sequence of application of HACCP is given in Figure 8.1.

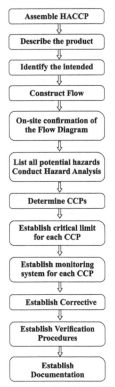

**FIGURE 8.1**   Sequence of steps in the application of HACCP.

## 8.3  RISK ANALYSIS

Risk is a function of the probability of an adverse health effect and the severity of that effect, consequential to hazard(s) in food. Risk analysis comprises three components[9] as indicated in Figure 8.2.

**FIGURE 8.2**   Three basic components in risk analysis.

Risk analysis is an iterative, ongoing and highly interactive (internal and external communication) process that should be evaluated and reviewed as necessary on the basis of new scientific data, information or changes in the context in which the food safety problem has occurred. It is a structured decision-making process.

It should be done based on scietific data available, and should be consistently reviewed. It should be an open, transparent process, documented completely. It also considers uncertainity and variability factors. It provides information and helps in decision making of food safety regulators. The essential conditions required for sucessful risk analysis are:

• Operational food safety system.
• Knowledge about risk analysis.
• Support and participation of key stakeholders.

Risk managers play a imprtant role in communication both internally and externally among stakeholders.[8]

In case of HACCP, hazard analysis is a part of risk assessment. During this process, the important considerations are:[1]

- Source of hazard.
- Probability of occurrence of hazard (Fig. 8.3).
- The nature of hazard.
- Severity of the adverse health effect caused by the hazard.

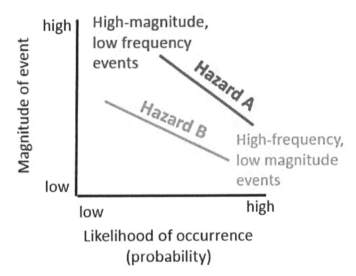

**FIGURE 8.3**   Probability of occurrence of hazard.

Based on the risk potential, it can be considered as CCP, OPRP, PRP, or GMP. In dairy industry, there are various *critical control points* (CPP) that are listed in Table 8.1.

## 8.4   ASSESSMENT OF DAIRY EQUIPMENT HYGIENE

It is necessary to examine the efficiency of cleaning and sanitation equipments by visual, chemical, and microbiological culture methods. In microbiological culture methods, commonly used methods to determine the surface contamination are swab and rinse methods. In case of swab

**TABLE 8.1** Critical Control Points in Dairy Industry.

| Product | Processing step | Hazard | Description | Critical limit/preventive measures | Reference |
|---|---|---|---|---|---|
| All dairy products | Reception of milk | Microbiological | Milk with high microbial load distribution under unhygienic conditions | Routine control determination pH(6.4–6.6)/acidity temperature control and vehicle cleaning | [2,11,14] |
| | | Physical | Foreign matter, dead insects, hair, and other materials | Filtration and microscopic examination | |
| | | Chemical | Antibiotic residues, pesticide residues, Aflatoxin M1, nitrates, and sanitizers | Tests for antibiotics and Aflatoxin M1, good veterinary practices, awareness programs to milk producers. Checks for residues. | |
| | Cold storage/distribution | Microbiological | multiplication of pathogen/spoilage organisms | Temperature control <6°C at cold storage and during distribution | |
| All Pasteurized dairy products | Pasteurization | Microbiological | Survival of Potential pathogen | Pasteurization temperature not < 72.5°C/15 s (or a similar time/temp profile), negative alkaline phosphatase test | [6] |
| UHT milk, sterilized cream | Sterilization | Microbiological | Render the product microbiologically sterile | 121°C/10 min (25% fat cream) 135°C/1 s (UHT Milk) | [15,16] |
| Milk, cream, ice cream | Cooling | Microbiological | Growth of potential pathogen/spoilage micro-organisms | Check temp/time profile of cooling post-pasteurization (2–4°C) | [22] |
| | Filling | Microbiological | Improper sealing of pouch/carton | Check by leakage test (drop test/squeeze test) | |
| | Labeling | Physical | Incorrect labeling | Check use before dates | |

**TABLE 8.1** *(Continued)*

| Product | Processing step | Hazard | Description | Critical limit/preventive measures | Reference |
|---|---|---|---|---|---|
| Fermented milk products | Starter culture inoculation/ fermentation | Microbiological | Potential pathogen survival | Check type/quantity or purity of culture check ripening time | [17] |
|  |  |  | Cross contamination | Check for correct storage Conditions of starter culture and hygiene status. |  |
| All types of cheeses | Addition of salt | Microbiological | Survival of potential pathogen | Check amount of salt used | [20] |

method, a swab is made-up of non-absorbent cotton, used for swabbing the equipment surface at defined area, whose hygiene is to be checked. In rinse method, the metallic cans, bottles, pipelines, etc., are rinsed with diluents like Ringer's solution and this diluent is analyzed by pour plate method. These are conventional methods that are time-consuming. The interpretation of results for swab, bottles, small containers, butter churns, and milk cans is done based on parameters that were defined by Motarjemi et al.[10]

The rapid method to detect microbial cells is faster than above conventional culture method, that is, ATP bioluminescence method. It measures amount of light generated by luciferase reaction with ATP molecules using luminometer. The ATPs present can be estimated within few seconds. The results are expressed in relative light units (RLU), which is directly proportional to the quantity of ATP.

## 8.5   REGULATORY REQUIREMENTS FOR LAYOUT AND DESIGN OF A DAIRY PLANT

### 8.5.1   PLANT LAYOUT REQUIREMENTS

#### 8.5.1.1   SITE

The dairy plant should be situated in a clean, pollution, and dust free environment away from cattle sheds, garbage dumps, open sewage drainages, and landfills. The surroundings should be free from obnoxious fumes, odors, excessive dust, and smoke.

#### 8.5.1.2   BUILDING

The structure shall be of suitable size, design, and construction to facilitate maintenance of hygienic operations for processing purpose. The material of construction shall be brick, plaster, cement, concrete, vitrified tiles, granite, and any other equivalent materials, which helps in ensuring hygiene. The building shall be pest-proof, provide sufficient space for equipments, storage of raw materials, packing materials and finished goods. No portion of the building should be used for domestic purposes like cooking food and staying unless it is separated physically with separate drainage facilities or separated from building itself.

## 8.5.1.3   VENTILATION AND LIGHTING

Depending on the nature of work, working duration (hours) and number of workers, adequate lighting and ventilation must be provided. The lighting fixtures should not be directly mounted on the process building, instead should be mounted on separate poles, focus to the building must be provided, to prevent attraction of insects. The tube lights and bulbs should be covered properly, to prevent glass breakage inside the processing area. Cross-ventilation needs to be provided for fresh air circulation, and optimum RH and temperature needs to be maintained in closed areas by air conditioning.

## 8.5.1.4   FLOORS, WALLS, AND CEILINGS

The floor should be washable, impervious, and smooth with appropriate slope to drains, to prevent any water accumulation after washing. In the CIP sections and processing areas, the floor should be covered with acid or alkali-resistant tiles to prevent any pitting due to corrosion. In storerooms, warehouses, floors should be hard, strong, and impervious in order to withstand movement of trolleys and forklifts. The internal walls and ceilings should have smooth, non-absorbent, light-colored surface. Must be free from crevices, and sharp edges, to facilitate effective cleaning. The corners or junctures between wall and floor or wall and wall should be rounded to prescribed radius to prevent accumulation of dust while proper cleaning.[19]

## 8.5.1.5   MAINTENANCE

The building should be maintained at good condition, must be dust free, and clean. At regular intervals, it may be washed, painted, disinfected, and treated for pest control. The paints with anti-mold agents are suitable for walls to prevent mold growth.

### 8.5.1.6   PROCESSING ROOM

The area should be fly-proof, rodent-proof, and insect-proof. It should not be open directly to open or garden area. The floor should be impervious, with chemical resistant tiles, should possess proper slope to drains. The drains should be kept clean, with correct slope to prevent, choking, water accumulation, or blockage. Drains should be covered with detachable covers to facilitate easy cleaning. The dead end of drain should be covered with screen to prevent entry of rodents and cockroaches. Adequate facilities, like number of water points, hot water, and steam points need to be provided to help in easy washing of the area. The exhaust fans to be installed in order to prevent condensation, and dripping problems.

### 8.5.1.7   WINDOWS AND DOORS

The processing area should be provided with self-closing doors, with air-curtains, or any means to prevent insect flying into the area. The whole area should be maintained with positive air pressure, to prevent flying insects and dust. The windows should be covered with fly-proof wire gauge or screen and should open outward.

### 8.5.1.8   DRAINAGE

The proper drainage should be provided to facilitate easy flow of large quantity of water used for cleaning. It is necessary to provide correct slope to prevent any water logging, or choking. Drain openings to be provided with screen traps to prevent solid matter from clogging.

### 8.5.1.9   INSTALLATION OF EQUIPMENTS

Any equipment should be installed on concrete foundation of durable and easily cleanable material. The equipment shall be placed at least 45 cm from wall, ceiling or sealed watertight thereto. It should be placed with sufficient clearance from wall, allowed to permit inspection, cleaning, and maintenance. Where pipelines pass through ceilings into the above floor, it should be spaced 5 cm above the floor, for proper cleaning.

### 8.5.1.10  CONNECTIONS

All electrical connections, switch boxes, panel boxes, conduit, and cables shall be installed at least 45 cm away from the equipment, walls to facilitate cleaning and maintenance.

## 8.5.2  EQUIPMENT AND STORAGE TANKS

- Instruments and working equipments intended to come into direct contact with raw materials and dairy products which are made of corrosion-resistant material (SS 304,316) and which are easy to clean and disinfect.
- Equipment, containers, and installations which come into contact with dairy products or perishable raw materials used during production shall be cleaned and if necessary disinfected according to a verified and documented cleaning program.
- The processing establishment shall in principle be cleaned according to an established, verified, and documented food safety management program. The manufacturer shall take appropriate measures to avoid any kind of cross contamination.
- Disinfectants and similar substances used shall be in such a way that they do not have any adverse effects on the machinery, equipment, raw materials, and dairy products kept at the dairy establishment. They shall be in clearly identifiable containers bearing labels with instructions for their use and their use shall be followed by thorough rinsing of such instruments and working equipments with potable water, unless supplier's instructions indicate otherwise.
- Any container or tank used for transporting or storage of raw milk shall be cleaned and disinfected before reuse.

The various design aspects to be considered in plant layout and equipment design are given in Table 8.2.

**TABLE 8.2**   Various Design Aspects to be Considered in Dairy Plant and Equipment Design.

| Plant/equipment design | Incorrect design | Correct design |
|---|---|---|
| Drainage pipes | | |
| Self draining tanks and vessels | | |
| Internal angles of wall of processing area | SQUARE CORNER | RADIUS |
| Permanent welded joints | Lap weld | Butt weld |
| Corner welds in equipment/ storage tanks | | |

## 8.6    BENEFITS OF FSMS FOR DAIRY BUSINESS

Food and waterborne diseases are leading causes of illness and death in developing countries; killing 2.2 million people annually, most of them being children. In food industry, effective implementation of FSMS can reduce the burden of business by avoiding recalls of the product due to hazards in the food, not meeting of regulatory requirements, and minimize the food-borne outbreaks.

### Benefits

- The FSMS provides the assurance about product quality, safety and reliability to consumers.
- It confirms the products to international quality standards; thereby helping the business to outperform the competitors in the market.
- It increases the profit margin, lets better utilization of resources, reduce product losses, reworks, efficient validation, and documentation procedures.
- Its valid base for decision-making, increased due diligence.
- Demonstrates the conformances to product quality as per regulatory requirements.

## 8.7    CONCLUSIONS

Guidelines requiring the application of the HACCP approach in dairy operations are significant, but will not be satisfactory to avoid the incidence of food-borne diseases without food operators having a thorough understanding of the concept and knowledge about safety issues. Food industry must give training to food handlers, which is a part of effective hazard analysis system. Training and education of regulatory officials in the principles and application of HACCP will also equip them better to undertake their role of verifying that HACCP plans developed by food businesses are correctly implemented and is effective. Finally, since most food businesses have limited understanding of HACCP and the procedures to implement it, it is necessary that every regulatory authority clarifies the goals of the strategy, and provides effective education and information to ensure uniformity in the application of its principles.

## KEYWORDS

- **Deming's cycle**
- **fermented milk products**
- **food safety**
- **hazard analysis**
- **hygiene**
- **quality assurance**
- **risk assessment**
- **safety assessment**
- **sanitation**
- **UHT processing of milk**

## REFERENCES

1. Akkerman, R.; Farahani, P.; Grunow, M. Quality, Safety and Sustainability in Food Distribution: A Review of Quantitative Operations Management Approaches and Challenges. *OR Spectrum.* **2010,** *32* (4), 863–904.
2. Arvanitoyannis, I. S. *HACCP and ISO 22000: Application to Foods of Animal Origin;* John Wiley & Sons: Hoboken, NJ, 2009; p 231.
3. Cattle and Dairy Development Fact Sheet Online. www.Dahd.Nic.In/about-us/divisions/cattle-and-dairy-development. (Accessed August 1, 2017).
4. Commission, C. A. *Recommended International Code of Practice-General Principles of Food Hygiene* (cac/rcp 1-1969, rev. 2 (1985)).*Codex Alimentarius.*1995, 1.1–2.0.
5. Harrigan, W. F.; Mccance, M. E. *Laboratory Methods in Food and Dairy Microbiology;* Academia Leon: Spain, 1979; p 231.
6. Kailasapathy, K. Chemical Composition, Physical and Functional Properties of Milk and Milk Ingredients. Chapter 4; *Dairy Processing and Quality Assurance*; John Wiley & Sons Inc.: New York; 2009; p 75–103.
7. Karaman, A. D. F.; Cobanoglu, R.; Tunalioglu, G. Barriers and Benefits of the Implementation of Food Safety Management Systems Among the Turkish Dairy Industry: A Case Study. *Food Control.* **2012,** *25* (2), 732–739.
8. Kuiper, H. A.; Davies, H. V. Are Safe Foods Risk Analysis Framework Suitable for GMOS? *Food Control.* **2010,** *21* (12), 1662–1676.
9. Mcmeekin, T.; Baranyi, J.; Bowman, P.; Dalgaard, M.; Kirk, T. Ross, S. Information Systems in Food Safety Management. *Int. J. Food Microbiol.* **2006,** *112* (3), 181–194.

10. Motarjemi, Y.; Käferstein, F. Food Safety, Hazard Analysis and Critical Control Point and the Increase in Foodborne Diseases: A Paradox? *Food Control.* **1999,** *10* (4), 325–333.

11. Nada, S.; Ilija, D. Implication of Food Safety Measures on Microbiological Quality of Raw and Pasteurized Milk. *Food Control.* **2012,** *25* (2), 728–731.

12. Noordhuizen, J.; Metz, J. Quality Control on Dairy Farms with Emphasis on Public Health, Food Safety, Animal Health and Welfare. *Lives. Prod. Sci.* **2005,** *94* (1), 51–59.

13. Oliver, S. P.; Jayarao, B. M. Foodborne Pathogens in Milk and the Dairy Farm Environment: Food Safety and Public Health Implications. *Foodborne Pathog. Dis.* **2005,** *2* (2), 115–129.

14. Oliver, S. P.; Boor, K. J. Food Safety Hazards Associated with Consumption of Raw Milk. *Foodborne Pathog. Dis.* **2009,** *6* (7), 793–806.

15. Papademas, P.; Bintsis, T. Food Safety Management Systems (FSMS) in the Dairy Industry: A Review. *Int. J. Dairy Technol.* **2010,** *63* (4), 489–503.

16. Robinson, R. K. *Dairy Microbiology Handbook: The Microbiology of Milk and Milk Products*; John Wiley & Sons: Hoboken, NJ, 2005; p 321.

17. Ruegg, P. Practical Food Safety Interventions for Dairy Production. *J. Dairy Sci.* **2003,** *86,* E1–E9.

18. Sharma, M.; Anand, S. Biofilms Evaluation as an Essential Component of HACCP for Food/Dairy Processing Industry—A Case. *Food Control.* **2002,** *13* (6), 469–477.

19. Tamime, A. Y. *Milk Processing and Quality Management*; John Wiley & Sons: Hoboken, NJ, 2009; p 112.

20. Tenenhaus-Aziza, F.; Daudin, J. J. Risk-Based Approach for Microbiological Food Safety Management in the Dairy Industry: The Case of Listeria Monocytogenes in Soft Cheese Made from Pasteurized Milk. *Risk Anal.* **2014,** *34* (1), 56–74.

21. Trienekens, J.; Zuurbier, P. Quality and Safety Standards in the Food Industry, Developments and Challenges. *Int. J. Prod. Econ.* **2008,** *113* (1), 107–122.

22. Wilbey, R. A. Microbiology of Cream and Butter. In *Dairy Microbiology Handbook,* 3rd ed.; Robinson, R. K. Ed.; Wiley: New York, 2002; pp 123–174.

# CHAPTER 9

# ECONOMIC ANALYSIS OF MILK MARKETING: CASE STUDY IN HARYANA

ANIL K. GUPTA*

*Department of Agricultural Economics, College of Agriculture, CCS Haryana Agricultural University, Hisar 125004, Haryana, India*

*\*E-mail: doc@agdentalclinic.com*

## CONTENTS

Edited and abbreviated version of *A. K. Gupta, (1993). An Economic Analysis of Milk Marketing In Haryana. Unpublished PhD (Economics) Dissertation submitted to the College of Agriculture, CCS Haryana Agricultural University, Hisar, Haryana, India.*

In this chapter: US$ 1 = 60.00 Indian Rupees (INR).

## ABSTRACT

The research study in this chapter aimed to: examine milk production, its consumption and disposal pattern in different categories of farms, study the factors affecting marketed surplus of milk, price behavior and price structure of milk in Haryana, and to work out the comparative efficiency of the private and cooperative sector milk plants.

This study revealed that the number of milch animals in a household increased with the increase in the size of land holding, thereby increasing the total daily milk production with increase in the size of farm. On an average, a farmer was producing 17.67 liters of milk per day. The average milk produced on a large producer–seller farm was almost three times more than the small milk producers. On an average, 31.07% of the total milk production on the selected farms was retained in the family for consumption in fluid form or after converting it into milk products. Average daily per capita consumption of milk on the selected farms was 646 g, which was about 20% more than the estimated daily milk availability of 539 g in Haryana, including all sections of the population. The per capita daily consumption of milk varied from 487 g for the small milk producers to 763 g on the large farms.

The marketed surplus of milk was 68.93% of the total quantity of milk produced on an average farm. The percentage of marketed surplus to the total milk production did not indicate any relationship with the size of producer–sellers. Functional analysis showed that all the explanatory variables included in the study explained more than 90% variation in marketed surplus of milk. Regression coefficients of per capita milk production $(X_1)$ and number of animals with milk $(X_4)$ were found to have positive and significant effects on marketed surplus of milk on all the categories of households as well as for pooled data. However, the coefficient for per capita milk consumption $(X_2)$ variable had a negative influence in all the three categories. The coefficient of price of milk per liter $(X_3)$ was positive for all the farm sizes, but not significant for large farm category. This indicated that the price of milk per liter did not influence much the marketed surplus of milk on large farms.

The study further showed that the average milk price increased throughout the period from the year 1979 to 1991 in all the three selected districts of Haryana due to increase in the general price level. The milk prices were abnormally high in the year 1987 as it was a drought year. The coefficient of variation (CV) giving month to month variation in prices of milk within each year was relatively higher in 1987 in all the districts. The milk prices ruled the highest in Jind among selected districts. This district also experienced maximum relative increase in milk prices due to lower milk production and supply in Jind district.

The monthly price indices of milk obtained through ratio to moving average method were relatively higher during summer months and lower during winter months. The trends and growth rates in milk prices estimated through exponential form showed wide variations across various districts. The highest annual compound growth rate (CGR) of 9.21% in milk price was observed in Jind, followed by Kurukshetra (8.87%) and Faridabad (8.54%). The highest CGR in Jind district is due to lower milk production, which is further due to inadequate availability of feeds and fodder.

The results of the study on the structure of milk prices indicated that on the whole, out of the total marketed surplus, 39.25, 41.95, 13.63, and 5.17% were channelized through milk vendors, milk plant, halwai, and consumers, respectively. There was direct relationship between marketing costs and the length of the channels. The most efficient channel was the producer–consumers in which both the producers and the consumers were benefitted.

Milk plants had the highest marketing cost of 259.64 Rs./100 kg of milk, while the marketing cost was minimum (Rs.94.36/100 kg) in channel I (producer–milk vendor–consumer). It was found that both the selected milk plants in the cooperative sector were using less than 50% of their installed capacity, while the milk plant in private sector was using about 72% of its installed capacity. From the cost structure of the milk plants, it is clear that the milk plant in private sector was spending more on processing of milk as compared to milk plants in cooperative sector, because of manu-facturing of high value products. The break-even analysis showed that the break-even quantity for Pehowa, Jind, and Ballabgarh milk plants were 43.675, 17.717, and 20.013 million kg of milk per annum.

## 9.1  INTRODUCTION

Indian agriculture continues to be dominated by the belief that its foun-dation is crop production. Its importance is beyond dispute, since food grains fulfill the first basic need by providing calories for sustenance of the growing population. Having achieved a level of self-sufficiency in cereal production, the base of farming needs to be broadened to enhance the quality in the daily diet, as represented by animal proteins. Milk is the most widely accepted and used animal product. The National Commission on Agriculture (NCA) also observed that, next to agriculture, dairy is the most important subsidiary occupation. In India, more than 80% of the cattle population is in rural areas and about 76% of the rural population is contributing toward milk production.

The dairy industry of India has indeed witnessed significant and enviable progress during the last four decades with the inception of the Operation Flood Program (OFP) in July, 1970. The first phase of OFP was aimed at capturing liquid milk market in four metropolitan cities by linking 27 milk sheds. During 1980–1985, the second phase of OFP, the program was extended to all the states in the country. As a result, about 34,500 dairy cooperative societies had been organized under 136 milk sheds. During 1985–1994, the third phase of OFP aims primarily at consolidating the extensive milk procurement and marketing network built during the second phase. Over 60,000 dairy cooperative societies have been organized in 173 milk sheds involving over 6 million members to procure an average of 8.3 million kg/day during April–September, 1989 as against 6.7 million kg in April–September, 1988 and marketed 7.3 million kg of milk a day in over 535 cities and towns of the country.[18–20]

A nationwide network of multi-tier milk producer cooperatives, democratic in structure, and professionally managed has come into existence. The institutional network comprising 22 federations, 173 unions, 60.8 thousand dairy cooperative societies and 7.0 million farmer members during 1990–1991, collected on an average 10 million liters of milk in a day and paid about 12 billion Indian rupees (INR) in a year. The milk procurement prices paid by cooperatives have increased at an annual rate of about 11% between 1980 and 1990. Milk producers now earn more income due to stabilization of milk prices between lean and flush seasons and creation of infrastructure of dairy cooperatives and milk collection centers for wider and interior areas. A milk grid has been established to balance seasonal and regional imbalances in milk production. Milk marketing has been expanded to supply hygienic and fair price milk to some 300 million consumers in 550 cities and towns. Dairy equipment manufacturing has grown to such an extent that most of the industry's needs are met indigenously. All these improvements could well be attributed to the intervention made by OFP—the largest dairy development project in the world.

Before the launching of OFP, India's annual milk production was static around 20 million tons but is now estimated to be well over 55 million tons. There was marginal increase in the milk production from 17.4 to 20.4 million tons (17.24%) during 1951–1961 and from 20.4 to 22.5 million tons (9.31%) during 1961–1971. The milk production increased annually by 0.451 million tons till 1951 while the annual rate of increase in milk production since 1965 has been 0.833 million tons. But after 1970's, dairy development started gaining momentum in the country through the implementation of various dairy cattle and buffalo improvement programs initiated by central government, state government, and nongovernment agencies (NGOs). With the results of these initiatives, the milk production showed an increasing trend. The level of milk production increased from 22.5 in 1971–1972 to 33.0 million tons in 1981–1982. During this period, there was 46.66% increase as compared to 9.31% increase during 1961–1971.

During 1961–1991, the milch animal productivity has increased from 688 to 972 grams per animal per day. There is no doubt that the milk production potential of Indian cattle is many times less than that of advanced countries. The average annual milk yield of Indian cow is 611 kg as compared to 7487 kg in Israel, 5510 kg in the USA, 5000 kg in Korea, 4829 kg in UK, 4543 kg in Japan, etc.

India, which used to import milk products, today ranks second in the world in total milk production, next only to the USA. The number of dairy plants has multiplied. Processing of milk which was less than a million liter a day is now over 12 million liters a day. Milk powder production has gone up from some 20,000 tons annually to over 165,000 tons. The need for import, even in the form of food aid, has been eliminated and a small beginning has been made to export milk powder.

Although the existing dairy development programs in the country have increased the milk production, but with the rise of human population, the increase in per capita availability of milk is only marginal. The per capita availability of milk wilted from 132 g in 1951 to 108 g in 1966. Later on after 1966, it increased from 108 to 112 g in 1971 and further from 1971 to 1981, the per capita availability of milk increased from 112 to 131 g (19.96%). The per capita milk availability was only 178 g per day during

1991–1992. India's per capita milk consumption is not commensurate with its ranking as the world's second largest milk producer. The minimum quantum of milk consumption recommended by the Nutritional Advisory Committee of the Indian Council of Medical Research (NACICMR) is 220 g/day (80 kg/year).[14]

Haryana state is not only the granary of India, but also possesses a high milk potential in the country due to its many favorable resource endowments like fertile land, assured irrigation, high yielding breeds of milch animals and above all, a receptive farming community with a proven record of early adoption of improved crop-milk technology. Haryana is at the 10th position with 5.8% of country's total milk production. The per capita milk availability is 539 g/day, which is second highest in the country only next to Punjab.

In Haryana, dairy cooperatives have a three-tier organizational structure. At the village level, there are dairy cooperative societies of milk producers which are federated into milk unions at the district level. At the state level, milk union has been federated into an apex body called Haryana Dairy Development Cooperative Federation Ltd. (HDDCF). The HDDCF is alone responsible for selling the entire milk collected by the dairy cooperatives in the state and milk products prepared at its dairy plants. There are 3257 milk cooperative societies in Haryana. These societies procured about 39.3 million liters of milk during 1990–1991. As a result of expansion of cooperatives, the milk production increased by 3.15 million tons during the same period in the past. Due to lack of transportation and proper technology for preservation, marketable surplus milk was confined to the villages, where it was converted into products or at the most, supplied to the nearby towns. Only less perishable commodities like ghee and khoa found the way out from the point of production. With the development of transportation, the milk now is transported as far as 350 miles or more for distribution in the metropolitan cities.

With the advent of technological changes, rapid growth of industries, fast increasing population in urban areas, increasing education, rising income, and nutrition consciousness by way of animal protein, the demand for milk and milk products being income elastic, is bound to increase in the coming years. Moreover, the latest technological breakthrough in milk production in terms of high yielding breeds of milch animals, nutritional varieties of fodders, better veterinary facilities, and establishment of semen banks, offer better prospects for milk products. These changing socioeconomic environments are adding new dimensions to various milk marketing problems like that of marketable surplus and consumption on one hand and marketing cost and margins in different marketing channels, milk price structure, and marketing practices on the other hand. Thus, the situation warrants such problems to be identified for finding out the solutions. The study in this chapter is an endeavor in this direction.

Milk being a highly perishable commodity, producers cannot retain it with them even for a short period and it needs a quick and an efficient marketing system. The existing marketing system in India is very often viewed as exploitive. There is a growing concern that the consumers are losing purchasing power due to rising prices, while the producers are not getting remunerative prices for their products. One of the main reasons advocated is the higher magnitude of the gross marketing margins, which is not only an additional burden on the consumers, but also disincentive to the poor producers. An efficient marketing is one which minimizes the cost of marketing services so as to ensure the largest share of the consumer's price to the producer. It does not however imply that marketing margin is the only criterion for efficiency. The milk consumer should also be assured of best possible services that is, good, clean, and wholesome milk and milk products at reasonable prices, so that the milk producer will have the incentive to increase production.

Most of the producers, being ignorant of the marketing methods, are generally cheated by the intermediaries and are paid a very low price. The collection and sale of rurally produced milk is dominated by hierarchy of mobile milk vendors. These middlemen not only exploit the producers by paying low prices and using faulty weights and measures, but also fleece the consumers by way of charging high prices and resorting to malpractices like under weighing and adulteration of milk with water and various other extraneous matters. Thus, the middlemen through their clever maneuverings, reap the maximum profit while producer–farmers being unorganized, having no provision to retain milk, are compelled to sell just after milking, at throw away (un-remunerative) prices.

Dairy cooperatives have been considered as one of the most important measures to improve the production and marketable surplus of milk on small dairy farms. The main objectives of milk producer cooperative societies are to safeguard and protect the interests of milk producers, organize marketing facilities for member's milk and fetch remunerative prices for milk for them. Being responsive to the needs of milk producers, it arranges to make available to the milk producers, the inputs like fodder seeds, cattle feeds, breeding, and veterinary facilities, to augment the milk production. Low production and inadequate marketable surplus of milk have been affecting the economy of the dairy plants. Dairy cooperative links of the milk producers in the rural areas and the consumers in the distant urban areas. Its efficiency is essential to achieve the goal set for the dairy industry as an instrument of economic and social change.

Estimation of marketable surplus of milk is important not only in determining the location of milk collection-cum-chilling centers and the size of milk plant before it is established, but its importance continues even after that for planning and projecting various development activities in the dairy sector. While in case of crop production, particularly food grains, it is now well established that bigger farmers are contributing more significantly toward the marketed surplus, and the same may or may not hold true for milk production. There is at present, a dearth of empirical evidence on the relationship of milk marketed surplus and various factors responsible for its variation.

With the increase in demand for milk in the urban areas and concentration of milk production activity in the rural areas, a large number of cooperative and private milk marketing agencies have entered into the business of procurement, processing, and distribution of milk. These milk marketing agencies were established with two major objectives: (1) to provide remunerative prices to the milk producer–sellers and (2) to ensure regular supply of milk to the urban consumer at affordable prices.

To achieve these objectives, precise and detailed information is required on the cost incurred on different operations in handling milk by different milk marketing agencies. This study, therefore, was undertaken to examine the role of organized sector in marketing and processing of milk and comparing efficiency of the alternative milk marketing agencies in Haryana. Specifically, the objectives of the study were:

- To examine the milk production, its consumption, and disposal pattern in different categories of farms;
- To study the factors affecting marketed surplus of milk;
- To study the price behavior and price structure of milk in Haryana; and
- To work out the comparative efficiency of the private and cooperative sector milk plants.

## 9.2   REVIEW OF LITERATURE

### 9.2.1   MILK PRODUCTION, ITS CONSUMPTION, AND DISPOSAL PATTERN

Ram et al.[36] in their study on marketed surplus of milk on farms found that the per capita consumption of milk was increased with an increase in land holding, but marketed surplus had inverse relationship with the size of land holding. The per capita consumption of milk was the lowest (300 g) in case of landless, which was followed by medium (468 g) and large (603 g) producers, respectively. This showed that landless and other small milk producers contributed more in increasing marketed surplus.

Ram and Solanki[37] studied the milk market structure in Karnal city of Haryana. The study revealed that there were three important milk marketing channels in the study area, namely, channel I (producer–milk vendor–consumer), channel II (producer–milk vendor–halwai/cream-eries–consumer), and channel III (producer (urban)–consumer). The producer's share in consumer's rupee was found much lower for village milk producers as compared to the urban producers. The producer's share in the consumer rupee was decreased with the increase in number of inter-mediaries in the marketing channels.

Thakur[50] also examined the extent of milk production and disposal by various categories of milk producers. It was seen that milk production per animal and the total marketed surplus of milk were higher in the case of the landless and small farmers as compared to the medium and large farmers. The landless farmers were found to market 76% of their produce. The landless and small farmers accounted for a large proportion of total marketed surplus of milk as compared to medium and large farmers.

Singh[42] in a study on production and marketed surplus of milk found that medium and large farmers were producing 26 and 27% of the total milk production in the study area. But their marketed surplus accounted for 20 and 18%, respectively, of total milk marketed. On the other hand, the share in the production of landless workers and small farmers were 23 and 24%, and their share worked out to be 29 and 32%, respectively. The study suggested that the main emphasis should be in favor of small farmers and landless workers for developing the milk market.

Singh et al.[46,48] conducted a survey to study the consumption pattern of milk and milk products in Vijayawada. The results revealed that 11% of the total income per month was spent on milk and milk products, particularly, fluid milk, and ghee. The per capita expenditure on milk and milk products was increased with income level.

Arputharaj et al.[3] examined the market structure of milk in Madras city and its price spread. The study revealed that in case of direct sales and sales through middleman, the producer gets a good price, but the consumer has to pay a high price without any guarantee about the quality of milk. In case of the Tamil Nadu Dairy Development Corporation (TNDC), the prices received by the producers are linked to the quality of his milk supplied. But in case of TNDC, consumer had to pay low prices per kg of milk.

Bhadure et al. (1981) in their study on milk production, consumption, and marketed surplus, reported a positive correlation between milk production, consumption, and size of holding that is, per household per day production and consumption of milk increased with the size of the operational holding. The production of milk per household was the lowest (5.03 L) in the landless category, however, the percentage of marketed surplus to milk production in this category was the highest (66%) followed by large (64%), and small farms (63%). Impact of milk production on marketed surplus of milk was found to be positively significant in all the categories of households.

Balishter et al.[5] studied the marketing of milk in Agra district. They reported that the producer's share in consumer's rupee was 67% in the first zone (village having distance of 15 km onward from the city boundary) in the entire herd size group. In all the herd sizes, the maximum amount of milk (90%) was sold through channel IV (producer–consumer). Producer–milk vender–consumer marketing channel was found to be efficient, because in this channel, the producer's share was maximum and consumer's price was also reasonable.

Barik et al.[6] pointed out that marketed surplus was decreased with the increase in size of holding. On the other hand, the quantity consumed increased with the increase in size of land holdings. They further reported that, as number of intermediaries increases, the percentage share of producer in the final price decreases. Khatik et al.[26] examined the disposal of milk by various categories of milk producers. It was noticed that generally the landless, small sized members marketed more quantity of milk, due to urgent financial needs.

Singh et al.[40] studied the production and disposal pattern of milk in rural Punjab. They reported that total quantity of milk produced on small, medium, and large holdings in the state were 27.77, 23.55, and 17.25% of total milk produced on three holdings. Quantity of milk consumed at home and converted into other products such as butter, ghee, etc., was 44.99 and 33.98%, respectively, of the total milk produced on an average farm holding. It was observed that the percentage of milk consumed at home and converted into products decreased with the increase in farm size, while percentage of quantity of milk sold increased with the increase in farm size.

Vashisht[51] observed that the producer share in consumer's rupee was the highest (92%) in Kangra milk supply scheme and the lowest (58.62%) under channel II (producer–wholesaler–retailer–consumer). However, the producers still preferred to sell their produce through the middlemen because of the services rendered by them and the inefficiencies of the organized sector.

Gill et al.[16] reported a positive correlation between farm size and the number of milch animals in Punjab. Although, milk production also varied directly with farm size, yet small farmers because of their low absolute requirements were able to have more marketed surplus than medium and large farmers. The family consumption had the first claim on milk supplies and the residual formed the marketed surplus. It was observed that the milk producers had a much higher level of consumption (850 g) per head per day, as compared with the minimum nutritional requirement of 210 g per head per day recommended by NACICMR.

Kumar et al.[27] reported that the share of marketed surplus and marketable surplus of milk in total milk production were 77.22 and 83.40%, respectively. In the case of landless and small dairy farmers, however, the marketed surplus exceeded marketable surplus. Patel et al.[33] reported that the milk produced per household showed a positive relationship with farm size. About 52% of total milk produced was consumed either directly, or after converting it into some product, and 48% of milk was sold to different agencies.

Singh et al.[39] studied the production and marketing of milk in Punjab. Depending upon the involvement of various agencies in the marketing of milk from the producer to the consumer, five marketing channels were observed. These were: channel I (producer–milk vendor–consumer), channel II (producer–milk vendor–halwai–consumer), channel III (producer–milkfed–consumer), channel IV (producer–halwai–consumer), and channel IV (producer–consumer). The channel III ranked first both in respect of the quantity of milk marketed and the number of farmers selling milk through it. The price spread ranged from 33 to 40% under the major channels of marketing of milk.[41,44]

Prabakaran et al.[35] examined the marketed surplus on various categories of farms. It was seen that per capita consumption of milk was the highest on medium sized farms. With the increase in farm size, milk retention also increased. In absolute terms, maximum milk was sold by large farms, whereas in percentage terms, share of the landless category was higher. Singh[40] estimated that marketable surplus on an average farm was 75% of the total quantity produced. The marketable surplus in urban areas (79.03%) was higher than in rural areas (62.20%). It was further seen that the producers' share in consumer's rupee ranged between 60 and 100%. Milk vendors shared about 19% of consumer's rupee in channel II (producer–milk vendor–halwai–consumer) and about 36% in channel IV (producer–halwai–consumer). The share of milkfed was 25% of final price. The price spread analysis indicated that producer's share in consumer's rupee decreased as the number of intermediaries increased in the marketing process. Kaur[23,24] reported that the major agency which was having the lion's share in the consumer rupee was the milk vendor followed by halwai and creameries. The producer's share in consumer's rupee ranged from 59.62 to 100%, depending on the length of the channel. Similarly, the share of the milk vendor was estimated to be 27.67, 12.67, and 12.35%, when milk was sold to consumer, creamery, and halwai, respectively. Milkfed retained 29.79% of consumer's price as their net margin.

Kainth[22] studied the milk marketing efficiency of different channels in Amritsar district of Punjab. The study revealed that marketing efficiency depended on various channels of distribution, which were affected by marketing margins and costs. There was a direct relationship between marketing costs and the length of the channels. The most efficient channel was found to be the producer–consumer in which both the producer and the consumer were benefitted in terms of total effectiveness, as it was the shortest one. The producer's share in the consumer's rupee was found to be 98.10% under this marketing channel. From the consumer's angle, the next important one was the producer–milk vendor–consumer channel. In this channel, out of the price paid by the consumer, 74.46% went to the producers, while 12.96% was the marketing cost and 12.62% was the net margin of the intermediaries.

Patel et al.[33] analyzed the marketed surplus of milk and various factors influencing it in different categories of households. The study revealed that marketed surplus formed about 49% of the total milk production and amongst different groups of households, the landless workers marketed the highest proportion (51%) followed by upper medium farmers (49%) of their milk produce over the year. Marketed surplus of the large farmers was the minimum (45.60%). Among the factors affecting the marketed surplus, milk production was found to be the most important for all the classes of households in all the seasons, while most of the households had no preference for selling milk to the organized sector. They reported that there was no significant distress sale of milk in the study area. Even among the poorest group of households, it was just marginal.

Gangwar et al.[13] analyzed the milk production and consumption in Haryana state. The study revealed that 62% of the total milk produced on the farms in the state was consumed by the family itself and 38% was sold as milk and milk products. Small farmers sold about 48% of the milk production, whereas medium and large farms sold only 39 and 36%, respectively. Per capita per day consumption was 491 mL on the small farms, 640 mL on medium farms, and 884 mL on the large farms. The study further revealed that prices of milk received by the farmers were same in the whole state. Net returns per liter of milk was highest (Rs. 0.52) on the small farms followed by medium (Rs. 0.46), and large farms (Rs. 0.09).

Kaur et al.[25] reported that per capita per day milk consumption was 620, 994, and 1128 g on small, medium, and large farms, respectively. It was seen that the marketed surplus of milk increased with the increase in milk production on different size of holdings, and the data for small, medium, and large farms were 5.70, 11.75, and 24.88 L, respectively. This constituted 64.71, 71.52, and 78.97 % of the milk produced on the respective farms. Singh et al.[45] studied the impact of dairy cooperatives on marketed surplus of milk in Western Uttar Pradesh. The study revealed that production as well as consumption of milk had no relationship with farm size. The overall per capita per day availability of milk was found to be 1190 g on the sample households as compared to 147 g in Uttar Pradesh. The marketed surplus of milk as percentage to total production was the lowest on large farms (60%) and the highest on medium farms (75%).

## 9.2.2 PRICE BEHAVIOR OF MILK AND MILK PRODUCTS

Mangat et al.[31] analyzed the seasonality in prices of milk based on data obtained from Ludhiana city in Punjab, milk shed area of the Moga Food Specialties Limited and milk shed area of Chandigarh milk supply scheme. Secondary data on monthly wholesale prices for the year 1960–1970 were collected from the Punjab State Statistical Organization. Seasonal price indices were worked out by using the time-series analysis. During the months when production indices ruled higher, price indices went low and vice versa. Thus, the producers were selling larger volumes of milk at low prices.

Kumbhare et al.[28] studied the weekly wholesale prices of buffalo milk for Bombay, Calcutta, Madras, Delhi, Patna, and Nagpur and ghee prices for the former four markets and butter prices for Bombay and Calcutta for the period January 1965 to December 1977 on the basis of data collected from the bulletin of agricultural prices, issued by the Directorate of Economics and Statistics—GOI, New Delhi. Monthly market prices derived from an unweighted averaging of weekly prices were used for analyzing the price behavior. The results of harmonic analysis confirmed the existence of seasonality in milk prices of Bombay and Delhi markets. The results based on ratio to moving average method evidenced that the seasonal variation were higher in Bombay compared to Calcutta and Delhi milk markets. Except Madras, the off seasonal increments in ghee prices were not discernible, possibly due to stocking effects and the high degree of market integration. Arora et al.[1] while studying the seasonal variation in the prices of milk, milk products, and feed constituents for the major metropolitan cities in India, namely, Bombay, Delhi, Calcutta, Madras, and Kanpur based on time-series data from 1963 to 1984 found the prices of milk in Delhi to be at its peak in June and showed declining trend up to December and showed increasing trend till it reached its peak. In case of ghee, the variations were less marked as compared to milk in all the markets. In Calcutta market, the price index varied from 99.26 (March) to 101.0 (September). In other markets too, the price index was maximum in July and at Calcutta market was maximum in May. It started declining and continued to decrease till July and thereafter, it showed an increase up to October and again showed decrease up to December. It showed an increase in January and February and was the lowest in March. The prices again increased in April. In

Kanpur market, the price indices was a little over than 100 from July to November, whereas, it was slightly less than100 during the remaining months. In Madras market, the index was below 100 from February to July and above 100 for the remaining period. But in Bombay market, the indices varied from 82 in December to 117 in July. However, the price indices were high from March to August.

Singh et al.[47] conducted time-series analysis in order to isolate the trend, seasonal, cyclical, and irregular fluctuations in milk procurement and milk prices. Data obtained from Karnal District Cooperative Milk Producers' Union for the period January, 1983 to March, 1988 formed the basis of study. Since, changes in milk procurement price have a direct bearing on the quantum of milk procured, the monthly variations in milk prices were also studied with the help of milk price seasonal index. Milk price indices showed narrow variations than the procurement indices. The index of milk price was highest during May, June, and July and it was least in February. Thus, the prices were higher by 3% over the average during May, June, July, and it went down by 6% in February. During March to August, October, and November the index was 100 and above while in the remaining months it was below 100. The price offered per kg of fat grew from Rs. 37 in 1983 to Rs. 56 in the year 1987.

Malhan et al.[30] observed that annual growth rates in milk and ghee prices ranged between 6 and 10% amongst various districts of Haryana. The highest annual compound growth rate (CGR) of 10.08% in milk prices was observed in Faridabad followed by Rohtak (7.92%) and Gurgaon (7.44%) and was minimum in Sirsa (6.00%). The per annum CGR in ghee prices ranged between 9.12% (Faridabad) and 6.12% (Kurukshetra). The study suggested that the milk prices would be highest in Faridabad while ghee prices in Sirsa. Arora et al.[2] studied the price movements of the milk and milk products. The study revealed a high degree positive relationship between prices of milk and milk products. The change per month in milk prices ranged from Rs. 1.17 in Delhi to about Rs.1.68 in the Calcutta market. The per month change in the prices of a quintal butter was observed to be Rs.13.29 and Rs.10.38 for Calcutta and Bombay market, respectively. The correlation coefficient between the prices of milk and butter varied from 94 to 98% in the various markets.

## 9.2.3   ECONOMICS OF MILK PROCESSING

Chaudhuri[7] conducted a study on pricing policy for fluid milk plants and marketing cost in the public sector milk plants. It was found that public sector milk plants suffered from heavy financial losses due to the under-utilization of plant resulting from inadequate milk procurement. This attributed to imbalance in the procurement and selling price for similar quality of product compared to market price. The low selling price stands in the way of giving a remunerative price to producers due to heavy financial losses. It was suggested that price of milk should be adjusted in accordance with the market forces.

Gruebele[17] observed that the labor productivity increased and per unit cost of labor in general decreased as plant volumes increased and as the number of items processed decreased. By limiting the number of items and by increasing plant volume, perhaps by merger, it would be possible to increase labor productivity or reduce per unit cost of labor in fluid milk plants. Labor productivity could also be enhanced by planning production more carefully, adopting new technology, doing a better job of motivating plant personnel, and keeping better records.

Garg et al.[15] investigated the problems faced by the cooperative milk board, Kanpur. They reported that total cost/100 L of standard milk borne by the milk board was Rs. 79.98. It included procurement cost (Rs. 15.17), processing cost (Rs. 24.13), administrative cost (Rs. 24.60), and selling and distribution cost (Rs. 16.08). The higher cost of processing was due to under-utilization of milk processing plant. They reported that mismanagement, misleading, and over expenditure was some of the important factors which decreased the efficiency of the processing organization.

Kunwar et al.[29] conducted a study to work out the economics of milk processing per liter under the public and cooperative units. The study revealed that the processing cost per liter in case of public unit was about one and half times higher than the cooperative society. The processing cost per liter of milk for the private and cooperative units was calculated at 36 and 24 paise, respectively. The higher cost in the public unit was due to low quantity of milk handled during the year. The fixed cost and maintenance cost per liter of milk buffalo in respect to cooperative and private farmers were calculated at Rs. 400, 1510, 600, and 1520, respectively. The gross income per milk buffalo per annum for cooperative farmers was Rs. 835 while in case of private farmers it was Rs. 189 only.

Somasekhara[49] reported that fixed cost constituted about 16% of the total cost of milk processing and the rest were the variable cost. Another study by Madhanon (1978) brought out that planning and scheduling play a vital role in the success and profitability of dairy organization. A well planned network of cooperative, supported by remunerative milk prices and input for increasing milk production would ensure adequate quantity of milk supply to the dairy plant all the year round. This in turn helped to avoid under-utilization of plants. Pawar et al.[34] compared the efficiency of private, public, and cooperative milk processing agencies in Western Maharashtra. The private agency was comparatively more efficient in terms of the cost of its rendering procurement, processing, and transportation and distribution services. The comparison of the total quantity of milk handled and installed capacity of milk processing plants revealed that the government cooperative and private agencies could utilize, respectively, 85.53, 76.00, and 97.04% of the installed capacity of the processing plants during the year. It was observed that milk producer's share in the consumer's price remained almost same (80%) in case of all the three milk marketing agencies. The profit margins of government, cooperative, and private agencies formed 2.25, 1.43, and 5.47% of the price of processed milk, respectively.

Sharma et al.[38] examined the cost of procurement, reception and chilling of milk for a public sector milk plant situated in North Western region of India. It was observed that the cost of procurement per kg of milk was Rs. 0.47 of which 67% was on transportation (50%) and labor (17%). The cost of reception was about Rs. 0.08/kg of milk received and out of which more than 60% (Rs. 0.05) was spent on fat and SNF tests alone. Chilling cost was about Rs. 0.05/kg milk.

Hagvane et al.[21] estimated the cost of chilling and operational efficiency at milk chilling plant. They observed that chilling plant was operating much below its installed capacity during 1977–1979. The under-utilization of plant has increased the chilling cost. But it was interesting to note that the chilling plant was operating above the break-even quantity. It was noticed that plant would generate a higher rate of return on capital invested. It implies that for better operational efficiency and rate of return on its capital, the plant should generate higher annual cost flow than the present level. This would be achieved only if higher quantity of milk is handled at minimum possible man-power and at maximum managerial efficiency at all levels.

Gajja et al.[12] analyzed the cost structure of milk marketing in the cooperative sector in Western Rajasthan. The analysis revealed that the price paid to the producers has a large share in the total cost in the processing. The transportation cost showed a decreasing trend with an increase in the quantum of milk handled. Similarly, processing cost also showed a decreasing trend. The distribution of milk witnessed an increasing trend throughout the study period in the Jodhpur milk plant. The marketing margin had shown an increase in the Jodhpur milk plant in 1973–1974 and 1976–1977. While in the Bikaner milk plant, it declined throughout the study period. The break-even point was 6.078 million liters and 12.386 million liters for Jodhpur and Bikaner milk plant, respectively. It was also revealed that the procurement of milk was much below the installed capacity of both the milk plants.

Gajja[11] made an attempt to compare the cost of milk processing in arid zone of Rajasthan. He concluded that the cost of collection of milk has lion's share in the total cost in the milk processing. The cost of raw material showed increasing trend while that of transportation indicated a decreasing trend. Tile study further revealed that procurement of milk was increasing but at decreasing, rates much below the capacity of milk plant. Singh et al.[43] conducted a study in North Western India on economic aspects of milk collection in a public sector plant. The study revealed that the cost of collection of a liter of milk was Rs. 0.1064 varying from Rs. 0.0812 to 0.1369 on different routes. The per liter cost of milk collection had a negative relation with the quantity of milk handled. The cost of collection of a liter of milk was found to be Rs. 0.89 for centers collecting below 5000 L of milk. The cost was decreased to Rs. 0.15 when milk collection was increased to 7584 L. The cost was Rs. 0.0536 for those centers where the collection of milk exceeded 30,000 L. The cost of collection could be decreased considerably by increasing the collection of milk possibly by recognizing the collection centers or diverting cluster of milk centers to one collection center.

Namhata et al.[32] noticed that the Trichy Milk Union is yet to fully utilize the plant capacity. Besides, the business on dairy products other than fluid milk has not influenced the break-even business of the organization significantly. It has been suggested that the union should diversify its plant to supplement the profitability; this in turn will open new avenues for milk production and marketing. Dhaliwal et al.[9] observed an inverse relationship between quantity of milk procured and incremental procurement cost in milk plants. It implied that higher the quantity of milk processed, lower would be the incremental processing cost. In the procurement of the milk, Bhatinda milk plant was most efficient with its minimum incremental procurement cost (Rs. 0.28/L and Hoshiarpur plant was the least efficient with its highest incremental procurement cost (Rs. 0.47/L). In the processing of milk, Mohali plant was most efficient with its minimum incremental processing cost (Rs. 2.39/L) and Ludhiana plant on the other hand was the least efficient with its highest incremental cost (Rs. 3.23/L).

Chauhan et al.[8] examined the economics of various milk procurement systems. The study revealed that average cost of procurement was the highest for the village level centers followed by chilling centers. It was observed to be the least for Dudhia level center and was attributed to the lower quantity of milk handled. The percentage share of cost of transportation, collection, and chilling of milk for the plant as a whole was 42.53 and 5%, respectively. This cost structure varied from system to system. The total expenditure on a kilogram of milk for the milk plant as a whole was Rs. 3.08 which varied from 3.01 in case of contractors to Rs. 3.65 for the chilling centers.

The findings of this study on comparative efficiency of private and cooperative sector milk plants will be very useful on the policy issues related to the establishment of new milk plants in Haryana in both the sectors as well as in improving the efficiency of existing milk plants.


## 9.3   METHODOLOGY

Major components of the methodology employed involve criteria used for the selection of milk plants, selection of respondents and intermediaries, method of data collection, and description of the techniques followed in the analysis of data.


### 9.3.1   SELECTION OF MILK PLANTS

There were seven milk plants in operation during the course of this study in Haryana state, out of which six were in cooperative sector and one in private sector. Cooperative milk plants are situated at Ambala, Ballabgarh, Bhiwani, Jind, Hisar, and Rohtak in the state of Haryana. Private milk plant is at Pehowa. Two milk plants from cooperative sector situated at Ballabgarh and Jind were selected purposely (Ballabgarh being its nearness to metropolitan city while Jind being situated in the major breeding tract of milch animals of the state). The only milk plant in the private sector was also selected purposely to compare its efficiency with cooperative sector milk plants.

## 9.3.2 SELECTION OF COOPERATIVE SOCIETIES/COLLECTION CENTERS

In consultation with official of selected milk plants, a complete list of villages having milk cooperative societies/collection centers was prepared for each selected milk plants separately. Three villages were selected randomly from each selected milk plant situated at different distances from the milk plants, and thus formed a sample of nine villages. The final selection of the villages is shown in Table 9.1.

## 9.3.3 SELECTION OF MILK PRODUCER-SELLERS

A complete list of all the milk producer–sellers in the selected villages along with their number of animals in milk was prepared. These producer–sellers were classified into three categories (i.e., small, medium, and large) on the basis of number of animals in milk by using cumulative cube-root frequency method. Twenty percentage of the producer–sellers from each category were randomly selected proportionately to the number of households in different categories. Table 9.2 indicates the three categories of milk producer–sellers in the selected villages; the final selection of the producer–sellers from the different categories.

## 9.3.4 SELECTION OF INTERMEDIARIES

There are large number of intermediaries involved in the collection and distribution of milk from producers to ultimate consumers. The major agencies involved in the marketing of milk are: milk vendors, halwais, cooperative societies/collection centers, and creameries. Each agency and each channel has its own significance and limitations. Some producer–sellers prefer one agency and some prefer another agency. A sample of five milk vendors, two halwais, and one creamery was selected randomly from the each selected villages to examine the pattern of price spread, marketing cost, and marketing margin. Thus, a sample of 45 milk vendors, 18 halwais, and 9 creameries was filially selected from the selected villages.

**TABLE 9.1**   Category-wise Selection of Villages Having Milk Cooperative Societies/Collection Centers.

| Approximate distance from milk plant (km) | Location of the selected cooperative societies/collection centers in selected milk plants | | |
|---|---|---|---|
| | Ballabgarh | Jind | Pehowa |
| < 10 | Machgarh | Ramrai | Pehowa |
| 10–25 | Jawa | Rajthal | Ismailabad |
| > 25 | Bahin | Safidon | Jhansa |

**TABLE 9.2**   Category-wise Sample Selected in the Selected Villages.

| Category | Number of animals in milk for each category | Total number of milk producer–sellers | Number of producer–sellers selected |
|---|---|---|---|
| Small | 1–3 | 352 | 70 |
| Medium | 4–6 | 403 | 81 |
| Large | 7 and above | 264 | 53 |
| Total | | 1019 | 204 |

## 9.3.5   COLLECTION OF DATA

### 9.3.5.1   MILK PRODUCTION, CONSUMPTION, AND DISPOSAL PATTERN

The data concerning size of dairy herd, family size, lactation period, volume of milk produced, consumed at home, disposed-off through various marketing agencies, and price received from various sales outlets were collected from producer–sellers with the help of suitably structured and pretested schedules by holding personal interview.

### 9.3.5.2   PRICE SPREAD AND MARKETING MARGIN

The data from milk vendors, halwais, and creameries were collected relating to volume of milk handled, price paid, and various cost incurred in transacting their business and price received by them from the consumers.

### 9.3.5.3   PRICE BEHAVIOR

To examine the price behavior of milk in Haryana, the secondary data on weekly consumer retail prices of milk (mixed milk of cow and buffalo) for Kurukshetra (Pehowa), Jind, and Faridabad (Ballabgarh) districts of Haryana were collected from *Economic and Statistical Organization, Planning Department*, Haryana. The average weekly prices were based on quotation of retail prices of the prescribed quality and unit of each item collected from selected shops. Monthly prices derived from an unweighted averaging of weekly prices were used for analyzing the price behavior.

### 9.3.5.4   PRIVATE AND COOPERATIVE MILK PLANTS

The data pertained to the amount of milk procured along with the procurement expenses on one hand and amount of milk handled in the processing sections of the plants along with processing costs on the other were collected from the private and cooperative sector milk plants. The Information relating to cost incurred on transportation, distribution, packing, etc., were also collected from various records maintained by different sections of the plant.

Other necessary information like consumer price index number, price index number for milk, etc., were collected from various published and unpublished materials, research reports, and resource personals in the field of milk marketing.

## 9.3.6   ANALYSIS OF DATA

### 9.3.6.1   MILK PRODUCTION, CONSUMPTION, AND MARKETED SURPLUS

To examine the pattern of milk production, its consumption and marketed surplus on different sizes of milk producing units, and to generate information on other parameters, the partial budgeting was carried out.

### 9.3.6.2   FACTOR AFFECTING MARKETED SURPLUS

To study the effect of various factors on marketed surplus, multiple regression analysis was carried out. Multiple linear regression and Cob-Douglas functions were fitted as follows:

Multiple linear regression equation:

$$Y = b_0 + b_1 X_1 + b_2 X_2 + b_3 X_3 + b_4 X_4 + U \tag{9.1}$$

Cob-Douglas function:

$$Y = b_0 X_1{}^{b_1} X_2{}^{b_2} X_3{}^{b_3} X_4{}^{b_4} e^u \tag{9.2}$$

$$\text{Log } Y = b_0 + b_1 \log X_1 + b_2 \log X_2 + b_3 \log X_3 + b_4 \log X_4 + U, \tag{9.3}$$

where $Y$ = Marketed surplus of milk (L); $b_0$ = Intercept; $X_1$ = Per capita milk production (L); $X_2$ = Per capita milk consumption (L); $X_3$ = Price of milk per liter (Rs.); $X_4$ = Number of animals in milk; and $U$ = Error term.

### 9.3.6.3   TIME SERIES ANALYSIS

One of the important objectives of price analysis is to construct the price relatives/indices, which are usually used to measure monthly/seasonal, temporal, and spatial changes in prices. Time series analysis technique was used to isolate seasonal fluctuations on prices of milk. In the classical or traditional approach, it is assumed that there is a multiplicative relationship among the four components and as such the following multiplicative model of time series was used:

$$Y_t = T_t \times S_t \times C_t \times I_t, \tag{9.4}$$

where $Y_t$ = Price of milk during the month; $T_t$ = Secular trend which shows long-term growth or decay; $S_t$ = Seasonal fluctuations superimposed on the trend (these are generally short term fluctuations of a recurring pattern); $C_t$ = Cyclical fluctuations which are of oscillatory nature; and $I_t$ = Irregular fluctuations, with the period of oscillations being more than a year; $T_t$ shows smooth, regular, and long term movement of a series. One complete period is known as a cycle. These fluctuations, which are not accounted by any of the above, are either wholly unaccountable or caused by such unforeseen events such as wars, floods, draught, etc.

### 9.3.6.3.1 Seasonal Index

Index number of seasonal variations is used to compare short term periodic fluctuations like those between months in a year. Relative seasonal fluctuations were calculated after eliminating the trend, cyclic, and irregular fluctuations with the help of 12 months moving averages by considering the multiplicative relationship of prices.

### 9.3.6.3.2 Trend Estimation in Milk Prices

The scatter diagram of milk data showed the possibility of a linear function. Therefore, the following form of the function was used to estimate the trend in milk prices:

$$Y_t = a + bt, \tag{9.5}$$

where $Y_t$ = Average price of milk during the time period t; $t$ = Time period expressed in year; $a$ = Intercept; and $b$ = Regression coefficient.

The CGR of milk prices was also estimated by analyzing the 13 years milk price data by using the following model:

$$Y_t = AB^t, \tag{9.6}$$

where $Y_t$ = Average price of milk during the time period t; $A$ = Constant; $B = 1 + r$; $r$ = CGR; $t$ = Time as trend variable in years.

### 9.3.6.3.3 Projection of Prices

The projected estimates of prices of milk for future years were estimated using following formula:

$$P_t = P_0 [1 + (r/100)]^t \tag{9.7}$$

where $P_t$ = Price of milk for the year t; $P_0$ = Price of milk in the base year; $r$ = Growth rate in the prices of milk over the period under study; and $t$ = 1–5, and 9 depending upon the price projection required for the period.

### 9.3.6.4   COMPARATIVE EFFICIENCY OF ALTERNATIVE MILK MARKETING AGENCIES

The comparative efficiency of private and cooperative sector milk plants was assessed on the basis of costs and returns in respect of the business handled by them in competition with each other. Besides the cost criteria, the market performance of these milk marketing agencies was judged by estimating the rate of return on investment and break-even point in respect of the individual agencies. The rate of return on investment was computed by solving the following equation for "$i$":

$$C = R_1/(1 + i) + R_2/(1 + i)^2 + \text{-} \text{-} \text{-} \text{-} + R_n/(1 + i)^n, \qquad (9.8)$$

where $C$ = Investment in capital assets; $R_i$ = Annual income flow assumed to be constant and is estimated as gross receipts minus costs (excluding depreciation and interest on fixed investment chargeable during the year); $i$ = Rate of return on investment or marginal efficiency of investment; and $n$ = Productive life period of investment.

The break-even quantity of milk was estimated as:

$$Q = [TFC]/[(P-AVC)], \qquad (9.9)$$

where $Q$ = Break-even quantity of milk; $TFC$ = Total fixed cost; $P$ = Price per liter of processed milk; $AVC$ = Average working cost per liter of milk.

## 9.4   RESULTS AND DISCUSSION

### 9.4.1   MILK PRODUCTION, CONSUMPTION, AND MARKETED SURPLUS

#### 9.4.1.1   SIZE OF FARM HOLDING AND NUMBER OF ANIMALS IN MILK

Results revealed that average farm holding was 2.98, 7.12, and 10.50 acres on small, medium, and large sample households, respectively. In terms of family size on an average, there were 6.44 members on small, 9.29 on medium, and 10.00 on large farm households. Average number of animals in milk on small, medium, and large farms was 2.34, 4.71, and 8.47 units, respectively. It was interesting to note that farm size has positive correlation with family size and number of animals in milk. These results are in consonance with Bahadure et al.[4] and Kaur et al.[25]

### 9.4.1.2  MILK PRODUCTION, CONSUMPTION, AND MARKETED SURPLUS

Average daily milk production on small, medium, and large farms was worked out to be 9.77, 17.22, and. 28.81 L, respectively. On an average, the quantity of milk produced per day per farm was 17.67 L. The data reveals that as the size of farm increases, the total quantity of daily milk produced also increases due to more number of animals in milk on larger farms.

On an average, daily 12.18 L of milk per household was sold constituting 68.93% of the total quantity of milk produced. On small, medium, and large farms, the daily marketed surplus was 6.63, 11.09, and 21.18 L which formed 67.86, 64.40, and 73.52% of total milk produced on these respective farm sizes. It was observed that although the milk production per household was the lowest on small category, yet the percentage of marketed surplus to milk production in this category was more (67.86%) than the medium category (64.40%). Economic compulsions are the main reasons for the higher percentage of marketed surplus of milk on the small category than medium category. The percentage of marketed surplus to total milk production did not indicate any relationship with the size of household as was reported by Ram et al.[36,37] in their study in milk shed areas of Karnal.

It was concluded that per household per day production and consumption of milk were increased with the size of farm households, indicating thereby, a positive correlation between the milk production, consumption, and size of farm household. Average daily consumption of milk per household was 5.49 L which amounted to 31.07% of the daily milk production. Positive correlation between milk production and its consumption may be due to the fact that as more milk was available with large farm households, the consumption of milk and milk products increased due to their better economic conditions.

## 9.4.1.3  *PER CAPITA PRODUCTION AND CONSUMPTION OF MILK*

It is evident from the data that per capita milk production on small, medium, and large farm households was 1.52, 1.85, and 2.88 L, respectively and per capita per day milk consumption followed the same trend with respect to size of farm as 0.487, 0.660, and 0.763 L on small, medium, and large farm categories, respectively, of the total daily milk production. However, in terms of percentage, the daily per capita consumption was the highest on medium size of farm (35.67%) and the lowest on the large farms (26.49%). It has also been observed that the level of milk consumption per head per day was much higher as compared to the minimum nutritional requirement of 220 g recommended by NACICMR. It was further interesting to note that per capita per day average milk consumption of 0.646 L in farming community was about 20% more than the estimated daily milk availability of 0.539 L in Haryana including all sections of the population.[10] This shows that a large surplus of milk existed on farm holdings in the state which could be tapped by developing on efficient marketing system keeping in view the interests of milk producers and consumers in the state.

## 9.4.2  *FACTORS AFFECTING MARKETED SURPLUS OF MILK*

To study the effect of factors on the marketed surplus of milk, the data were subjected to functional analysis for each size of farms as well as pooled together. Both linear and Cob-Douglas type of functions were fitted. The equation giving the best fit and conforming to the production and economic logic were considered for further interpretation of results. The linear and Cob-Douglas type of functions were tried between marketed surplus of milk (Y) with other independent variables like per capita production ($X_1$), per capita consumption ($X_2$), price of milk per liter ($X_3$), and number of animals in milk ($X_4$).

Table 9.3 summarizes the results of the regression analysis to study the various factors influencing the marketed surplus of milk in respect of each of the farm. It can be observed that all the explanatory variables included in the study explained more than 90% variation in marketed surplus of milk. The variation was observed to be the highest on large farms (96.91%) and lowest on medium farms (92.12%). The marketed surplus function for the pooled data (all size of farms) explained about 95% variation.

TABLE 9.3    Marketed Surplus Function of Milk in Selected Districts of Haryana.

| Regression coefficient | Farm size | | | |
|---|---|---|---|---|
| | Small | Medium | Large | Pooled |
| Intercept ($b_0$) | 0.7857 | 6.0050 | 22.5141 | 6.9656 |
| Per capita production ($X_1$) | 0.0984** | 0.3367** | 0.4373** | 0.1079** |
| | (2.8658) | (8.0922) | (6.6677) | (2.6835) |
| Per capita consumption ($X_2$) | −0.1172** | −0.2297** | −0.4962* | −0.0775** |
| | (3.3627) | (5.3024) | (2.3242) | (2.7726) |
| Price of milk per liter ($X_3$) | 0.8233** | 0.5627* | $0.1089^{NS}$ | 0.5568* |
| | (7.1418) | (2.1207) | (1.1589) | (1.9930) |
| Number of animals in milk ($X_4$) | 0.1599** | 0.2892** | 0.8323** | 0.3657** |
| | (3.5869) | (4.6968) | (8.1369) | (9.4195) |
| Coefficient of multiple determination ($R^2$) | 0.9646 | 0.9212 | 0.9691 | 0.9515 |

It is interesting to note that the regression coefficient of per capita milk production ($X_1$) variable was positive and significant statistically at 1% level in all the categories of households as well as for pooled data. This indicated a significant impact of per capita production on marketed surplus of milk in all the categories of households. However, the coefficient of per capita milk consumption ($X_2$) variable was negative and significant at 1% significance level for small and medium farms indicating thereby a negative impact of this variable on the marketed surplus of milk. Coefficient of per capita milk consumption was significant at 5% level for large farm households. It could be further observed that coefficients of price of milk per liter ($X_3$) were positive for all the farm sizes but were not significant for large category, while they were significant at 1 and 5% level for small and medium farms, respectively. This indicated that increase in price of milk for small household would generate higher marketed surplus but this factor does not encourage the growth of marketed surplus on large farm households.

The regression coefficient of number of animals in milk ($X_4$), as expected, was significant at 1% level and positively influencing the marketed surplus of milk on each size of farm as well as for the pooled data.

## 9.4.3   PRICE BEHAVIOR AND PRICE STRUCTURE OF MILK

### 9.4.3.1   PRICE BEHAVIOR OF MILK

The milk production is seasonal in nature and the prices of the fluid milk and milk products vary inversely with the changes in production. An understanding of prices could be of immense help to the dairy industry in short- and long-term forecasting. Thus, it is important to study the behavior of milk prices.

#### 9.4.3.1.1   Seasonal Variation in the Price of Milk

One of the important objectives of the price analysis is to measure price changes from one time to another and also to compare prices among different places. Therefore, it becomes necessary to examine the price behavior of milk over years in selected districts. The prices for each of the 12 months in various years were subjected to analysis of variance to look into the variation among years and among months in each of the three selected districts of Haryana state separately. As there were marked differences in prices of milk across different districts of Haryana, therefore, mean prices of milk along with coefficient of variation (CV) were considered for each of the three selected districts. The seasonal variation in the monthly prices of milk was examined by studying the average prices of milk along with CV over years in different districts.

The CV values and data reveal that in general the average milk price increased from 1979 to 1991 in all the three districts of Haryana (Jind, Kurukshetra, and Faridabad). This may be attributed to the increase in the general price level. The highest average milk price of Rs. 6.91/L was observed in Jind district. Also this district showed the maximum increase in the milk prices. This may plausibly be due to lower milk production and supply in Jind district. The reason for such a lower milk production may be due to inadequate availability of feeds and fodders which cannot be grown due to arid conditions prevailing in this district.

### 9.4.3.1.2   Monthly/Seasonal Fluctuations in Price of Milk

The milk production is seasonal in nature and the prices of fluid milk vary inversely with changes in production. Therefore, it becomes essential to examine the seasonal variation in prices of milk. The monthly price index was estimated by *ratio to moving average method* in order to look into monthly/seasonal fluctuations in milk prices. The monthly indices were estimated for each of the three selected districts separately.

The monthly price indices for milk in Jind district gave wide variations. The indices ranged between 94.46 and 110.53. The maximum index of 110.53 was in the month of June reflecting a rise of 10.53% in prices over the average as 100. The minimum price index was in the month of January (94.46). The monthly price indices showed a declining trend from July onward till January whereafter, there was a rise in the index till the month of June.

In Kurukshetra district, the monthly price indices for milk ranged between 94.83 (January) and 106.91 (July) showing relatively narrow variation in monthly milk price indices. The range of monthly variation in the price index was approximately 6.00% on either side of the average price level which can be stated as approximately 6.00% above and below the average price level. The declining trend in monthly indices were noticed from the month of August onward and reached its trough in January whereafter the indices started showing an increasing trend in July.

The monthly price indices for Faridabad ranged between 95.83 (January) and 104.81 (July). This shows that the milk prices were higher in July by 4.81% and were lower by 4.17% in January over the average price level. Milk prices showed an increasing trend from January to April, then recorded a dip in the month of May and rose again till July. From August, the milk prices started receding and came down to minimum in January.

The results indicate that the monthly indices of milk were maximum during June–July in different districts implying thereby that the milk prices were higher during these months. These indices were observed to be minimum in January in all the districts suggesting that the prices were lowest during this month. The monthly indices were lower than hundred from November to February which constitute winter season in all the three districts. This suggests that the milk prices were lower during winter. In March, the monthly price index was again less than 100 in all the selected districts, thereafter, the monthly indices started rising and reached its peak somewhere between June and July implying that the milk prices reached their peak level during summer months in selected districts. From July onward, the price indices started to decline and as such the milk prices were lower in monsoon season as compared to summer season but definitely higher than the winter. These findings are in agreement with those of Kumbhare and Patel,[28] Arora and Singh,[1] and Singh and Verma.[47]

Above discussion on monthly price indices for milk leads us to conclude that the prices were relatively higher during summer months and lower during winter months in all the three selected districts of Haryana. The milk prices were observed to be at its peak between June and July and were lowest in January. The higher milk prices during summer months may be attributed to lower milk production and supply which may again be due to inadequate availability of feeds and fodder and scarcity of water and also most of the cows and buffaloes were in the advance stage of pregnancy. Relatively lower prices during winter and monsoon months may be due to better milk supply during these periods. Further, during these months, the cows and buffaloes are at their peak level of milk production due to better availability of feeds and fodder.

Relatively narrow variations were observed in the monthly price indices of milk in Kurukshetra and Faridabad as compared to Jind district. This may be attributed to better irrigation facilities in those districts and as such there is better availability of green fodder throughout the year which contributes to the better availability and supply of milk. On the other hand, high range in monthly price indices observed in Jind district may plausibly be due to dry and arid conditions which are not suitable for growing green fodder giving rise to uneven milk production and supply throughout the year.

### 9.4.3.1.3 Consumer Price Trend and Price Projection

Different agro-climatic conditions, breeds of cows and buffaloes, their genetic potential, feeds and fodder availability and marketing infrastructure lead to wide variations in milk production in different regions. These factors result in imbalance in demand and supply of milk, which ultimately affects the price structure of milk and milk products. Therefore, the trend in prices of milk shows wide variation in different regions. With this objective in mind, an attempt was made to estimate the growth in prices of milk for selected districts of Haryana. Time-series data were used for fitting various trend equations. Two algebraic forms of trend equations (linear and exponential) were tried to estimate the growth in price. The coefficient of multiple determination associated with each of the two algebraic forms were also estimated. The results obtained for both the forms were compared on the basis of economic as well as statistical criteria. A perusal of the data and further examination of results revealed that the *exponential form* gave the best results and therefore, it was selected for further study and estimation of growth rate.

## Price trend of milk

Table 9.4 presents the estimate of regression parameters based on *exponential form* and the annual CGR in milk prices for various selected districts of Haryana. The coefficient of determination ranged from 0.9243 for Faridabad district to 0.9868 for Kurukshetra district. Table 9.4 reveals that the regression coefficient was statistically significant ($p < 0.01$) in all the three selected districts. It was found that the highest growth rate of 9.21% was observed in Jind may plausibly be due to lower milk production which is further due to inadequate availability of feeds and fodder being dry and arid.

**TABLE 9.4**   Estimates of Regression Parameters and Annual Growth Rate in Milk Prices for Three Districts in Haryana.

| District | Constant | Regression coefficient | T value | Coefficient of determination ($r^2$) | Growth rate (%) |
|---|---|---|---|---|---|
| Jind | 2.2208 | 1.0921 | 23.3452 | 0.9802 | 9.21 |
| Kurukshetra | 2.1122 | 1.0887 | 28.6854 | 0.9868 | 8.87 |
| Faridabad | 2.5291 | 1.0854 | 11.5408 | 0.9243 | 8.54 |

## Price projection of milk

The projections of milk prices for the period 1992–1996 and 2000 were obtained for selected districts of Haryana taking into account the present trend in milk prices. The price projections were made assuming that the current rate of growth in milk prices would continue for selected districts.

Table 9.5 gives the projection of prices of milk in selected districts for the period 1992–1996 and 2000. A perusal of the Table reveals that the milk prices are expected to be Rs. 14.83/L in 2000 in Jind district, which will be highest milk price among all the three selected districts. The per liter milk prices are expected to be Rs. 13.97 in Kurukshetra and Rs. 13.32 in Faridabad in 2000. The projected milk prices indicated that with the present growth rate, the milk prices will increase by about 59.61% in Jind, 52.92% in Kurukshetra, and 48.43% in Faridabad in the year 1996 over 1991. The milk prices are expected to increase by more than 100% in 2000 over 1991.

**TABLE 9.5**   Projection of Milk Prices for Three Districts in Haryana (Rs./L).

| District | 1992 | 1993 | 1994 | 1995 | 1996 | 2000 | Percent increase in 1996 over 1991 | Percent increase in 2000 over 1991 |
|---|---|---|---|---|---|---|---|---|
| Jind | 7.33 | 8.00 | 8.74 | 9.54 | 10.42 | 14.83 | 59.61 | 121.01 |
| Kurukshetra | 7.08 | 7.70 | 8.39 | 9.13 | 9.94 | 13.97 | 52.92 | 114.92 |
| Faridabad | 6.91 | 7.50 | 8.14 | 8.84 | 9.60 | 13.32 | 48.43 | 109.10 |

## 9.4.3.2   STRUCTURE OF MILK PRICES

Marketable surplus of milk is disposed-off by the producers through different milk marketing channels. Market channel is sequence of different agencies through which a commodity passes till it reaches the ultimate consumer. The major milk marketing channels involved in the collection and distribution of milk in the study area were as follows:

Channel I        Producer–milk vendor–consumer

Channel II       Producer–milk vendor–creamery–consumer

Channel III      Producer–milk vendor–halwai–consumer

Channel IV       Producer–halwai–consumer

Channel V        Producer–milk plant–consumer

Channel VI       Producer–consumer

The importance of each agency was judged from the quantity of milk handled. The quantity handled by these agencies in the study area is presented in Table 9.6. It can be observed that small and medium farmers sell their maximum quantity of marketable surplus to the milk vendors. This may be due to the fact that milk vendors make advance payments to these categories of farmers.

The perusal of Table 9.6 shows that on small farm in the total quantity of marketed surplus, milk vendors, milk plants, halwais, and consumers constitute 48.11, 22.32, 8.45, and 21.12%, respectively. On medium farms, however, 42.65, 36.97, 16.86, and 3.52% of total marketed surplus was disposed-off through milk vendors, milk plant, halwai, and consumers, respectively. Large producers sold 32.81, 54.06, and 13.13% of the marketed surplus to the vendors, milk plants, and halwais, respectively. The inter-category comparison shows that medium category milk producers sold comparatively higher proportion of their surplus milk to halwais while the small farmers sold higher proportion direct to consumers.

**TABLE 9.6**  Patterns for the Disposal of Marketed Surplus on Different Size of Households in (L/day).

| Category | Average quantity of marketed surplus | Quantity of milk (%) sold to | | | |
|---|---|---|---|---|---|
| | | Milk vendor | Milk plant | Halwai | Consumer |
| Small | 6.63 | 3.19 | 1.48 | 0.56 | 1.40 |
| | (100.00) | (48.11) | (22.32) | (8.45) | (21.12) |
| Medium | 11.09 | 4.73 | 4.10 | 1.87 | 0.39 |
| | (100.00) | (42.65) | (36.97) | (16.86) | (3.52) |
| Large | 21.18 | 6.95 | 11.45 | 2.78 | – |
| | (100.00) | (32.81) | (54.06) | (13.13) | |
| Average | 12.18 | 4.78 | 5.11 | 1.66 | 0.63 |
| | (100.00) | (39.25) | (41.95) | (13.63) | (5.17) |

Data in parentheses indicate percentages to total marketed surplus.

The results also indicate that there was negative correlation among the proportion of milk sold to vendors and farm size, whereas in case of milk plants, the proportion of milk sold and farm size were found to be positively correlated. Direct sale to consumers was negligible because of less local demand as most of the village people kept milch animals and had sufficient quantity of milk to meet the daily milk requirement. Large milk producers did not sell milk to local consumers as they were generally not interested in retail sale.

On the whole, out of total marketed surplus, 39.25, 41.95, 13.63, and 5.17% was channelized through milk vendor, milk plant, halwai, and direct to consumers, respectively. There was, thus, a parallel co-existence of unorganized and organized milk market. The unorganized milk trade was mostly in the hands of milk vendors as they collected milk from farmer's door step. Only a small quantity of marketable surplus (13.63%) was disposed-off through halwai, which may be because it required a lot of time and efforts on the part of producer to supply milk to halwai shops situated at distant places from the village.

### 9.4.3.2.1   Price Spread in the Marketing of Milk

#### Producer's share

The data clearly indicate that producer's share in consumer rupee was maximum (100%) in channel VI due to direct sale to the consumer without incurring any marketing cost. However, it could not be much practiced due to negligible local demand for milk. From producer's point of view, next best channel was channel I where producer received 75.84% share of consumer's rupee. It was found that producer's share in consumers rupee was further decreased when halwai or creamery entered in marketing channel along with milk vendor. However, in absolute terms, the producer's price remained same (Rs. 582.43/100 kg) in channel I, channel II and channel III. Producer's share in absolute term was minimum in channel V, which may be due to low fat percentage in milk supplied to the milk plant.

#### Milk vendor's share

It has been observed that the milk vendors play an important role in supplying milk to the consumers, halwais, and creameries in the cities. Milk vendor's share was highest (24.16%) in channel I, whereas it was minimum (10.63%) in channel II. It may be because of the fact that milk vendors charged higher prices from them, while creameries paid less price to the milk vendor as they pay money to the milk vendor on fat basis.

## Milk plant's share

The data indicate that milk plant's share in consumer rupee was 30.22%. This agency handled about 41.95% of the total quantity of milk produced by the milk producers in the study area. The agency is known for making prompt payments to the producers after every 10 days.

## Halwai's share

As only a small portion (13.63) of the total marketed surplus was channelized through this agency, this agency is not of much importance. The share of halwai in consumer's rupee was very high and was 25.65 and 19.26% in channel IV and III, respectively. The share of this agency was higher than other agencies because the halwai charged much higher prices (Rs. 8.39/L) of milk from the consumer.

## Creamery's share

Creamery's share in the consumer's rupee was 18.26%. This agency has also been found responsible for adulteration of milk by milk vendors. As milk vendors get cream separated for about half of their milk from these creameries and sell the cream as such or take it to home for converting it into ghee. They, then, mix the second half of the milk with the separated milk and sell this mixed milk to the consumer at the whole milk price.

### 9.4.3.2.2   Marketing Costs and Margins of Various Agencies

Data indicate that costs incurred (Rs./100 kg) by producers were 35.85 and 26.34 when milk was sold to halwai and milk plant in channels IV and V, respectively. Accordingly, the costs incurred (Rs./100 kg) by milk plants and halwai were 233.30 and 94.35, which were higher than the costs incurred under any other channel. These higher costs may be on account of under-capacity utilization of the milk plant. halwai's costs were more because he incurred costs on account of shop rent, utensils, furniture, electricity bill, labor, and miscellaneous costs. Amongst the marketing costs incurred by milk vendor in channels I, II, and III, marketing costs was maximum (Rs. 94.36/100 kg of milk) in channel

I where he moved from house to house to distribute milk. In channels II and III, he sold milk to halwai and creamery in large volume and thus the cost was relatively low. Marketing costs borne by creamery was minimum (Rs. 57.99/100 kg), which may be due to large volume of business handled by them.

The marketing margins of various agencies under different channels of milk marketing were worked out by deducting marketing costs incurred by the intermediaries from the gross margin of these respective intermediaries. The data shows that milk vendor received a net margin of 91.21, 22.44, and 30.44 Rs./100 kg milk handled in channels I, II, and III, respectively. The margin was maximum when he sold milk to consumer at a higher price and minimum when he sold left out quantities to creamery. The results show that net share of milk vendor depends upon the type of channels adopted.

The halwai received a net margin of Rs. 67.25 (68.91%) and Rs. 120.85 (100.00%) under channels III and IV, respectively. The higher net margin (Rs. 120.85/100 kg) to halwai in channel IV was due to the fact that he paid less price to milk producer and charged maximum price from the consumer. Similarly, the net margin of creamery was also quite high (Rs. 91.51/100 kg of milk: 80.31%) on account of higher prices charged by them from the consumers. Under channel V, milk plants incurred losses during the year 1991–92 @ Rs. 10.00/100 kg of milk handled. This may be due to extreme under-utilization, inadequate-management, higher processing costs, fixed costs, etc.

From producer's point of view and on the whole, it may be concluded that channel VI (producer–consumer) was the most efficient yielding maximum price for his produce and it was followed by channel I (producer–milk vendor–consumer), channel IV (producer–halwai–consumer), channel II (producer–milk vendor–creamery–consumer), channel V (producer–milk plant–consumer), and channel III (producer–milk vendor–halwai–consumer).

From the consumer view point, the best channel was V (producer–milk plant–consumer) from where he get wholesome milk at the minimum price, followed by channels VI (producer–consumer), channel I (producer–milk vendor–consumer), and channel II (producer–milk vendor–creamery–consumer), whereas channel III (producer–milk vendor–halwai–consumer) and channel IV (producer–halwai–consumer) were at par.

## 9.4.4   COMPARATIVE EFFICIENCY OF PRIVATE AND COOPERATIVE MILK PLANTS

With the increase in the demand for milk in the urban areas and concentration of milk production activity in the rural areas, a large number of cooperative and private milk marketing agencies have entered into the business of procurement, processing and distribution of milk. During the past decades, the milk production activity has gained its importance in the rural areas of Haryana mainly because of remunerative prices for milk and the establishment of different milk marketing agencies engaged in the procurement of milk from the rural areas and distribution of processed milk to the consumers in the cities.

The cooperative and private traders compete with each other not only in procurement of raw milk from the producers but also in distribution of processed milk and milk products to the consumers. Because of perfectly competitive market for milk, these agencies are unable to make profits through manipulation of price either of raw milk or processed milk and, therefore, the only alternative left to them is to manage the business efficiently in such a way that the cost of their rendering procurement, processing, transportation, and distribution services remains low. Moreover, for the economic viability of the milk plants it is demanded that milk plants should not only meet its variable and fixed costs but it should also generate handsome net profits to introduce latest technology in the processing of milk. This section, therefore, has been designed to study the comparative efficiency of the private and cooperative milk plants in Haryana.

### 9.4.4.1   SIZE OF THE BUSINESS

The data shows that the installed capacity was maximum (109.5 million liters: 1991–1992) for private milk plant while both the cooperative milk plants had installed capacity of 36.5 million liters each. The total quantity of milk procured by these plants was 79.00, 16.932, and 15.478 million kg for Pehowa, Jind, and Ballabgarh milk plants, respectively. The milk plant in cooperative sector were using less than 50% of their installed capacity while private was using about 72% of its installed capacity. It is interesting to note that nearly whole of the procured milk was utilized for the manufacturing of high-value milk products in case of private milk plant. Cooperative milk plants sold major portion of their procured milk in the fluid form after standardization. The discrimination among the plants in the utilization pattern of milk procured determines their economic viability. The spoilage of milk formed 0.74–4.78% of the total quantity of milk procured, the lowest and highest being in case of private and cooperative milk plant (Jind), respectively. These wastages and losses of milk also affect the overall cost and revenue structure of the individual plants. These results show that the private milk plant can handle milk more efficiently than the cooperative sector milk plants.

### 9.4.4.2   COMPARATIVE COST STRUCTURE OF MILK-HANDLING OPERATIONS

#### 9.4.4.2.1   Raw Material

The total cost of the business was 836.35, 677.71, and 664.02 Rs./100 kg of the total quantity of milk procured in case of Pehowa, Jind, and Ballabgarh milk plants, respectively. Looking to the item-wise cost of business, it is observed that the cost on account of purchase of raw milk alone was 76.85, 72.74, and 79.08% of the total cost, respectively, in the case of Pehowa, Jind, and Ballabgarh milk plants; and the remaining cost was on account of labor, material supply, transport, agent's commission, etc.

#### 9.4.4.2.2   Procurement Cost

All the expenditure incurred on milk comprising price for the transportation, chilling charges, and commission was paid to societies engaged in milk, collection at various milk collection centers/ cooperative societies. The price was paid to the producers on the basis of fat and solid non-fat (SNF). The procurement cost was 20.24, 26.41, and 20.84 Rs./100 kg of milk procured for Pehowa, Jind, and Ballabgarh milk plants, respectively. The comparison of procurement cost of private and cooperative sector milk plants showed that cooperative milk plants spends more on procurement due to lower milk procurement compared to their milk handling capacities and procurement infrastructure available. Further, it was reported that Ballabgarh milk plant was procuring milk from short distances (less than 100 km), while Jind milk plant collects milk from far away villages. This had increased the procurement cost for Jind milk plant compared to Ballabgarh milk plant.

### 9.4.4.2.3  Processing Cost

After reaching milk to dairy plant either from milk collection centers or cooperative societies, milk goes under chilling to make standardized milk and milk products. The processing cost consists of electricity, water, chemical, furnace oil, labor charges, maintenance, depreciation, and interest. The processing cost was 90.67 and 71.47 Rs./100 kg of milk for Jind and Ballabgarh milk plants, respectively. The processing cost was higher in case of private milk plant, which was 100.70 Rs./100 kg of milk procured. Higher cost of processing in case of private milk plant may be due to the fact that they were using procured milk for manufacturing of high value products like ghee, SMP, etc.

### 9.4.4.2.4  Packing Cost

The packing cost is determined by the nature of product being manufactured by an individual milk plant. Data revealed that the cost of packing was highest in Jind milk plant (Rs. 52.50) followed by private milk plant at Pehowa (Rs. 48.93) and Ballabgarh milk plant (Rs. 26.93)/100 kg of milk. Jind and Pehowa milk plants utilized 61.20 and 99.20% of their procured milk in manufacturing of milk products, which require superior metallic containers and hence the cost of packing has been inflated. Whereas, Ballabgarh milk plant sold major portion of its procurement in the form of fluid standardized milk, which requires cheap packing material (i.e., polythene bags).

### 9.4.4.2.5  Distribution Cost

The distribution cost mainly includes advertisement charges, transportation, and commission paid to booth agents. It is essential to place the processed milk and other milk products in the market for disposal. In case of cooperative milk plants, the Ballabgarh milk plant incurred Rs. 19.67/100 kg of milk procured, which was followed by Jind with Rs. 15.16. The higher cost of distribution in case of Ballabgarh milk plant was mainly due to the fact that it daily supplies large quantity of standardized milk in Delhi and nearby areas, which has enhanced the distribution cost. The distribution cost in case of private milk plant was Rs. 23.75/100 kg of milk procured. The comparison of distribution cost of private and cooperative milk plants indicates that distribution cost was highest in case of private milk plant as compared to cooperative milk plants. This was primarily due to the huge expenditure on advertisement and sales promotion by private milk plant.

### 9.4.4.3  COMPARATIVE ECONOMICS OF MILK PROCESSING

The total cost of milk processing was 100.70, 90.67, and 77.45 Rs./100 kg of milk in Pehowa, Jind, and Ballabgarh milk plants, respectively. The component-wise cost analysis revealed that out of total processing cost, the administration charges alone accounted for 40.30, 42.10, and 45.89% in Pehowa, Jind, and Ballabgarh milk plants, respectively. Administration expenditure includes salary of staff engaged, stationery, travelling allowances, telephone, post and telegraph, medical, and other expenses. The data indicated that cooperative milk plants had created more infrastructure facilities as compared to private milk plant. Moreover, due to political interference, cooperative milk plant management is forced to employ more personals, which increase the administration expenditure of these milk plants. The situation is very much different in case of private milk plant, which was fully utilizing its installed capacity and employing trained/skilled and efficient workers.

The data shows that the private milk plant was spending more on power and fuel as compared to cooperative milk plants. In percentage term, the expenditure on power and fuel varied between 18.96 (Jind) and 24.68% (Pehowa). The expenditure on power and fuel was 20.13% for Ballabgarh milk Plant. It is also evident that expenditure on chemical was highest in case of private milk plant (11.47%) followed by Jind (9.10%), and Ballabgarh (7.81%).

The higher expenditure on power, fuel, and chemicals in case of private milk plant may be attributed to the product mix they manufacture. As the private milk plant is engaged in manufacturing of high value product like ghee, skimmed milk powder, butter, etc., which require comparatively more energy for longer period and a lot of chemicals. Hence, the cost of power, fuel, and chemical per unit of milk processed has increased in case of private milk plant. The maintenance and repair costs were comparatively higher in case of cooperative milk plants as compared to private milk plant. The repair and maintenance cost was 10.17, 10.85, and 9.77 Rs./100 kg of milk for Pehowa, Jind, and Ballabgarh milk plants. More maintenance and repair cost in case of cooperative milk plant may be attributed to mishandling of equipment and machinery by untrained workers employed by milk plant management.

The miscellaneous cost incurred by cooperative milk plant seems to be well in order because Jind milk plant had expanded its capacity recently, which pushed the depreciation and interest upward. The miscellaneous cost depends upon the nature of products manufactured, type of technology adopted, and period of establishment of plant. The foregoing discussion reveals that the more cost of processing in case of private milk plant does not mean that the plant is not optimally utilized as the cost of processing depends on the nature of products made, capacity utilization, type of technology, and human capital.

### 9.4.4.4   GROSS AND NET RETURN

The gross returns are total sale proceeds received on the sale of milk and milk products in a given period of time. The net return is total profit worked out by deducting the total cost from total revenue. The gross and net returns depend on the volume of business, price per unit of product, besides the type of products being manufactured by an individual milk plant. The gross return was 9.59, 7.49, and 7.24 Rs./kg in Pehowa, Jind, and Ballabgarh milk plants, respectively. The total working costs were 8.36, 6.78, and 6.64 Rs./kg in the respective milk plants. The fixed costs were highest (Rs. 0.77/kg) in Ballabgarh milk plant followed by Jind (Rs. 0.74/ kg) and Pehowa milk plant (Rs. 0.68/kg). Data shows the private agency would reap relatively the highest profits from the business. The per unit profit over total working cost in case of cooperative milk plants was lower mainly due to more losses on account of spoilage of milk and non-diversification of business for the preparation of milk products. Further, the profit over total cost was also higher in case of private milk plant as compared to cooperative milk plants which clearly indicate that private milk plant was utilizing the installed capacity optimally and working efficiently. The augment of procurement and diversification of milk plants will go a long way to bring the cooperative milk plants in profitable position.

## 9.4.4.5   BREAK-EVEN ANALYSIS

In short run, if total revenue covers the variable cost, a plant may continue in operation and otherwise not, that is, price equal to or higher than average working cost. In case of long run, the total revenue covers the fixed as well as working cost. Fixed cost includes depreciation on building, interest on capital investment. The break-even analysis indicates that private milk plant operated the business at a much higher level of break-even point while both the cooperative milk plants operated much below the break-even quantity.

The analysis indicated that the estimated break-even output of milk were 43.675, 17.717, and 20.013 million kg/annum at the prevailing average price in Pehowa, Jind, and Ballabgarh milk plants, respectively. The comparison of the total quantity of milk handled and installed capacity of milk processing plants revealed that milk plants situated at Pehowa, Jind, and Ballabgarh could utilize 72.15, 46.39, and 42.40% of the installed capacity (1991–1992). These variations in the utilization of installed capacity have necessarily influenced tile costs and returns structure of the business to a greater extent. Both the milk plants in cooperative sector were working below the break-even quantity and thus not covering even the cost of production. It may therefore, be suggested that the plant should increase its scale of operation through higher procurement of milk in order to cover at least its operational expenses.

## 9.4.4.6   RATE OF RETURN ON INVESTMENT

From the foregoing discussion regarding costs and returns structure of the private and cooperative milk plants, it is revealed that the private agency could derive relatively higher profits from the business because of the diversification of the business and economical and efficient management of the procurement, processing, transportation, and distribution services. On the score of per unit profitability, the cooperative milk plants were relatively less efficient in comparison with private milk plant. However, since the investment in capital assets varied considerably among the agencies, the comparison contemplated so far was not alone sufficient to judge the market performance of the individual agencies. The analysis was therefore, extended further to estimate the rate of return on investment to facilitate meaningful comparison of the efficiency of the alternative milk marketing agencies.

It is clear from the data that the rate of return on investment was the highest in case of private milk plant and lowest in the case of cooperative milk plant at Ballabgarh. Besides, the rate of return on investment exceeded the market rate of interest in the case of private milk plant even if the productive life period of investment is considered 10 years. In case of cooperative milk plants, however, the rate of return on investment was less than the market rate of interest assuming productive life of the investment of 10 years. It implies that for better operational efficiency and rate of return on capital in case of cooperative milk plants, the plants should generate a higher annual cash flow than the present level. This would be achieved only if higher quantity of milk is handled at minimum possible cost and at maximum managerial efficiency at all levels.

## 9.5  CONCLUSIONS AND POLICY IMPLICATIONS

- The farm size has positive correlation with family size and number of animals in milk. As the size farm increases, the total quantity of daily milk produced also increases due to more number of animals in milk on large farms.
- The milk production per household is the lowest on small-category; however, the percentage of marketed surplus to milk production in this category is more than the medium category. Economic compulsions are the main reasons for the more percentage of marketed surpluses of milk on the small category than medium category. The percentage of marketed surplus to total milk production does not indicate any relationship with the size of household.
- The production and consumption of milk increases with the size of milk producer-sellers, indicating thereby, a positive correlation between the milk production, consumption, and size of milk producer–sellers.
- The level of milk consumption per head per day is much higher as compared to the minimum nutritional requirement of 220 g recommended by NACICMR. Thus, a large surplus of milk exists on farm holdings in the state which could be tapped by developing an efficient marketing system keeping in view the interests of milk producers and consumers in the state.

- The increase in price of milk for small household generates higher marketed surplus but this factor does not encourage the growth of marketed surplus on large farm households.
- The milk prices are observed to be at its peak between June and July and were the lowest in January in all the three selected districts of Haryana.
- The milk plant in cooperative sector are using less than 50% of their installed capacity while private milk plant is using about 72% of its installed capacity. Nearly whole of the procured milk are utilized for the manufacturing of high value milk products in case of private milk plant. Cooperative milk plants sold major portion of their procured milk in the fluid form after standardization.
- The per unit profit over total working cost in case of cooperative milk plants is lower mainly due to more losses on account of spoilage of milk and non-diversification of business for the preparation of milk products.
- The plant should increase its scale of operation through higher procurement of milk so that the total costs of the plant are less than the total returns.
- The plants should generate a higher annual cash flow than the present level for better operational efficiency and rate of return on capital in case of cooperative milk plants. This would be achieved only if higher quantity of milk is handled at minimum possible cost and at maximum managerial efficiency at all levels.

## KEYWORDS

- **cost structure**
- **Haryana Agricultural University**
- **Indian Agriculture National Commission on Agriculture (NCA)**
- **marketing cost marketing network**
- **milch animal**
- **processing of milk**

## REFERENCES

1. Arora, V. K.; Singh, K. *Study of the Price Movements of Milk, Milk Products and Feed Ingredients in the Principal Metropolitan Cities of India;* Annual Report: NDRI, Karnal, Hariyana, India, 1986; pp 66–67.
2. Arora, V. K.; Singh, K. Price Movement of Milk and Milk Products in Different Markets. *Agric. Marketing.* **1991,** *34* (4), 33–36.
3. Arputharaj, C.; Rajagopalan. R. Market Structure of Milk in Madras City. *Indian J. Agric. Econ.* **1979,** *34* (3), 213.
4. Bahadure, J. Z.; Singh, C. B.; Patel, R. R. Milk Production, Consumption and Marketed Surplus in Some Villages around Karnal. *Agric. Marketing.* **1981,** *24* (3), 19–22.
5. Balishter, L.; Singh, D. A Study of Marketing of Milk in Agra District. *Agric. Marketing.* **1981,** *24* (3), 25–30.
6. Barik, B. B.; Singh, R. K. Production and Marketing of Milk. *Indian Dairyman.* **1981,** *33* (2), 77–82.
7. Chaudhuri, A. K. R. Pricing Policy for Fluid Milk Plant and Marketing. *Indian Dairyman* **1968,** *20* (6), 185–191.
8. Chauhan, A. K.; Arora, V. K.; Singh, K. Economic Analysis or Various Milk Procurement Systems—A Comparative Study. *Indian J. Animal Res.* **1990,** *25* (1), 29–35.
9. Dhaliwal, N. S.; Khattra, P. S. Cost-Volume Analysis of Dairy Plant. *Indian Dairyman* **1990,** *42* (6), 278–281.
10. Economic Advisor. Haryana; Statistical Abstracts of Haryana (1991–1992); Government of Haryana, Department of Statistics, Chandigarh, India.
11. Gajja, B. L. Cost Analysis of Milk Processing in Arid Zone of Rajasthan. *Indian Cooperative Rev.* **1985,** *23* (1), 83–93.
12. Gajja, B. L.; Vyas, D. L. Cost Structure of Milk Marketing in the Cooperative Sector in Western Rajasthan. *Indian J. Agric. Econ.* **1984,** *39* (3), 243–244.
13. Gangwar, A. C.; Panghal, B. S.; Kumar, K. An Economic Analysis of Milk Production and Consumption on Different Sizes of Farms in Haryana State. *Indian J. Dairy Sci.* **1989,** *42* (4), 676–683.
14. Government of India. Economic Survey (1991–1992); Department of Agricultural Statistics, Government of India, New Delhi, pp 35–40.
15. Garg, J. S.; Prasad, V. An Economic Investigation into the Problems of Cooperative Milk Board, Kanpur. *Indian J. Agric. Econ.* **1975,** *30* (3), 143–144.
16. Gill, G.; Patel, R. K. Milk Production and Disposal Pattern in the Rural Punjab. *Asian J. Dairy Res.* **1983,** *2* (1), 50–54.
17. Gruebele, J. M. Measure of Efficiency in Milk Plant Operation. *Illinois Agric. Econ.* **1973,** *13* (2), 28–43.
18. Gupta, J. N.; Patel, R. K. Marketed Surplus of Milk in Rural Karnal. *Asian J. Dairy Res.* **1988,** *7* (2), 97–107.
19. Gupta, J. P. Disposal Pattern of Milk in Punjab. *Indian J. Dairy Sci.* **1992,** *45* (6), 292–293.
20. Gupta, P. R. *Dairy India-Annual Book;* Government of India Publication: New Delhi, India, 1992; pp 10–40.

21. Hagvane, L. R.; Kulkarni, M. B.; Chavan, I. G. Cost of Chilling and Operational Efficiency at Milk Chilling Plant–A Case Study. *Indian Dairyman.* **1982,** *34* (11), 197–202.
22. Kainth, G. S. Efficiency of Milk Marketing Channels in Amritsar District of Punjab. *Asian J. Dairy Res.* **1988,** *7* (1), 39–44.
23. Kaur, B. Study on Milk Supply System in Ludhiana City of Ludhiana District. M.Sc. Thesis (Unpublished), Punjab Agricultural University, Ludhiana, 1987.
24. Kaur, M. Study on Marketing of Milk in Ludhiana District (Punjab). M.Sc. Thesis (Unpublished), Punjab Agricultural University, Ludhiana, 1987.
25. Kaur, M.; Gill, G. S. Milk Production, Consumption and Disposal Pattern in Rural Areas of Ludhiana District (Punjab). *Indian J. Dairy Sci.* **1989,** *42* (4), 689–693.
26. Khatik, I. N.; Kulkarni, M. B. Marketed and Marketable Surplus of Milk in Different Categories of Milk Producers. *Indian Cooperative Rev.* **1981,** *19* (1), 47–54.
27. Kumar, Y.; Arora, V. P. S. Changing Milk Marketing Organization–Direction and Consequences. *Indian J. Agric. Econ.* **1984,** *34* (3), 231.
28. Kumbhare, S. L.; Patel, R. K. Price Behavior of Dairy Products in Selected Market. *Indian J. Dairy Sci.* **1982,** *35* (4), 497–504.
29. Kunwar, R.; Chauhan, Y. S.; Singh, R. l. A Comparative Study of Milk Processing by Public and Cooperative Units in Kanpur District, Uttar Pradesh. *Indian J. Agric. Econ.* **1975,** *30* (3), 143.
30. Malhan, R. S.; Jain, D. K. Trends and Projections of Milk and Ghee Prices in Haryana. *Agric. Marketing.* **1990,** *32* (4), 30–33.
31. Mangat, H. S.; Gill, K. S. Structure of Milk Prices in Punjab. *Agric. Marketing.* **1974,** *24* (4), 17–24.
32. Namhata, D.; Broadway, A. C. Production and Sales of Dairy Products in Trichy District Cooperative Milk Producers Union, Tamil Nadu–A Break-even Analysis. *Indian Cooperative Rev.* **1988,** *25* (3), 307–315.
33. Patel, R. K.; Singh, C. B.; Sharma, S. P. Production, Consumption, and Marketed Surplus of Milk in DRP Villages. *Indian Dairyman.* 1984, *36* (7), 365.
34. Pawar, J. R.; Sawant, S. K. Comparative Efficiency of Alternative Milk Marketing Agencies in Western Maharashtra. *Indian J. Agric. Econ.* **1979,** *34* (4), 160–167.
35. Prabakaran, R.; Sivaselvan, S. N. A Study on Marketed Surplus and Supply Functions for Milk in Chengalpathu District of Tamil Nadu. *Indian J. Dairy Sci.* **1986,** *29* (1), 13–16.
36. Ram, K.; Singh, K.; Patel, R. K. Marketed Surplus of Milk on Farms–A Pilot Study. *Indian Dairyman* **1973,** *25* (8), 297–300.
37. Ram, K.; Solanki, R. S. Study of Milk Market Structure in Kamal City, Haryana. *Agric. Marketing.* **1975,** *18* (2), 8–12.
38. Sharma, S. K.; Singh, R. V. Cost Analysis of Procurement, Reception, and Chilling of Milk. *Indian J. Dairy Sci.* **1980,** *33* (1), 75–81.
39. Singh, B.; Bal, H. S.; Kumar, N. Production and Marketing of Milk in Rural Areas of Punjab. *Indian J. Agric. Econ.* **1984,** *39* (3), 224.
40. Singh, Bant; Singh, Rachhpal; Bal, H. S. Production and Pattern of Disposal of Milk in Rural Punjab. *Agric. Marketing.* **1981,** *24* (1), 21–26.
41. Singh, J. Study into the Production and Marketing of Milk in Ludhiana District (Punjab). M.Sc. Thesis (Unpublished), Punjab Agricultural University, Ludhiana, 1986.

42. Singh, K. Who Feeds the Milk Market? *Indian Dairyman* **1978,** *30* (4), 277–280.
43. Singh, K.; Ram, K. Cost of Collection of Milk in a Public Sector Plant. *Asian J. Dairy Res.* **1987,** *6* (3), 130–134.
44. Singh, Lotan; Singh, C. B. Impact of Dairy Cooperatives on Marketed Surplus of Milk in Western Uttar Pradesh. *Agric. Marketing.* **1989,** *32* (1), 9–12.
45. Singh, Raj Pal; Singh, K. Economics of Milk Production in Meerut District (U.P.). *Indian J. Dairy Res.* **1987,** *6* (3), 125–129.
46. Singh, R. V.; Patel, R. K.; Rao, N. J. M. Consumption Pattern of Milk and Milk Products. *XX Int. Dairy Congress*; Anand, Gujarat, India, **1978,** *E* 1051–1052.
47. Singh, S.; Verma, N. K. *Trends and Seasonal Fluctuations in Milk Procurement;* Unpublished Report, National Dairy Research Institute: Karnal, Haryana, India, 1988.
48. Singh, S. P.; Singh, P.; Tiwedi, C. B. An Economic Analysis of Interstate Disparities in Milk Production and Institutional Facilities in India. *Agric. Situation India.* **1986,** *40* (10), 889–894.
49. Somasekhara, N. Cost Components of Dairy Manufacturing Industry—A Case Study. *Agric. Situation India* **1975,** *30* (8), 575–579.
50. Thakur, D. S. Impact of Dairy Development through Milk Cooperatives—A Case Study of Gujarat. *Indian J. Agric. Econ.* **1975,** *30* (3), 183–189.
51. Vashisht, G. D. Dynamics of Milk Production and Marketing in Himachal Pradesh. Ph.D. Thesis (Unpublished), HPKVV, Palampur, India, 1981.

# CHAPTER 10

# DAIRY FOODS: ALLERGY AND INTOLERANCE

MOHAMMED NAYEEM[1], GURDEEP RATTU[2], ROHANT DHAKA[1], AJAY KUMAR KASHYAP[3], NISHANT KUMAR[3], and PRAMOD KUMAR[4*]

[1]Department of Food Science and Technology, National Institute of Food Technology Entrepreneurship and Management (NIFTEM), Kundli, Rai, Haryana, India.

[2]Department of Basic and Applied Science, National Institute of Food Technology Entrepreneurship and Management (NIFTEM), Kundli, Rai, Haryana, India.

[3]Department of Agricultural and Environmental Science, National Institute of Food Technology Entrepreneurship and Management (NIFTEM), Kundli, Rai, Haryana, India.

[4]Dairy Chemistry Division, National Dairy Research Institute, Karnal 132001, Haryana, India.

*Corresponding author. E-mail: param.bhu@gmail.com

## CONTENTS

## ABSTRACT

Molecular approach like genetic engineering is being explored. One of the approaches is to silence or modify the genes of specific allergenic proteins. However, genetic engineering is regarded as a complex process as milk contains lot of allergenic proteins, and each protein exhibits multiple allergenic epitopes. No food allergen has been removed completely by genetic methods but efforts are going on worldwide. This chapter presented concepts of lactose intolerance and milk allergy.

## 10.1 INTRODUCTION

Dairy is one of the most nutritive food items since time immemorial. However, there could be several reasons to avoid milk. It may be the inability of our body to digest the proteinaceous part of the milk or due to lactose, that is, milk sugar, which does not get digested and instigates allergic response. By comparing and contrasting, this chapter discusses two cases: (1) milk allergy and lactose intolerance and (2) technical advances achieved in overcoming such limitations of dairy foods.

### 10.1.1 LACTOSE INTOLERANCE

Nearly 20% population suffers symptoms of lactose intolerance and is more frequently observed in some ethnic groups. Lactose intolerance is not a case in which allergy is induced, rather it is a condition in which the body is unable to produce enough enzyme namely lactase, which digests lactose component of the milk and thus, causes problems for the body in terms of digestion. Both children and adults are affected by this, and symptoms usually observed are: diarrhea,[39] and bloating, etc. Lactose intolerance is a genetic disorder but can be caused by viral or bacterial infection in the small intestine. It can also happen for short period of time due to gastroenteritis but such cases get resolved on its own by healing ability. However, in persons born with primary lactose intolerance or those who develop it at a later stage, symptoms generally prevail throughout the life of the individual.[38]

Quantity of lactose in most of the domestic mammal milks (such as cows' milk, goats' milk, buffalo milk, and sheep's milk) is nearly equal. Effect may vary from person to person and also depends on dose, for instance, symptoms may not be observed in an individual who consumes a small amount of milk in tea, but the same individual may suffer serious problems related to indigestion on consumption of a glass of milk. Lactose intolerance has a global prevalence, which leads to common gastrointestinal disorders. Persons who are intolerant to lactose are advised to abstain from milk and other dairy products so as to avoid such disorders but such practice may cause deficiency of essential nutrients particularly that of calcium. Those, who are lactose mal-digester, can consume milk and dairy products without confronting symptoms by modifying their dietary pattern. Another methodology, which can be adopted by lactose intolerant people, includes replacement of milk with yogurt, fermented dairy products, soya milk, etc.

## 10.1.2  MILK ALLERGY

Milk allergy is often confused with lactose intolerance. Allergy due to consumption of milk is an overreaction of the immune system to certain proteins present in milk. The most common milk protein responsible for triggering allergic response is alpha-S1-casein.[20] When these proteins are ingested, it leads to an allergic response that may include symptoms like itching, hives, swelling, rashes, etc. In extreme cases, severe symptoms such as trouble while wheezing, breathing, and loss of consciousness, etc., can also be noticed. In fact, an acute allergic reaction to milk protein to which the body has become hypersensitive can lead to anaphylaxis thus, consumption of milk can be life threatening. The reason why allergic responses are not found in young persons on consumption of breast milk is that alpha-S1-caseins differ between species.[3]

Both the cases (i.e., lactose intolerance and milk allergy) are discussed in detail in this chapter.

## 10.2  LACTOSE DEFICIENCY AT A GLANCE

### 10.2.1  LACTOSE INTOLERANCE

Lactose is the pre-dominant and a unique carbohydrate that is found in milk and milk products. It is a disaccharide that is combination of galactose and glucose (Fig. 10.1). The main source of lactose is milk (cows, goat, and sheep) and dairy products (cream, cheese, butter, yogurt, cakes, and biscuits). Mature human milk contains 7.2 g/100 mL and cows milk contains 4.7 g/100 mL of lactose. The lactose formula is $C_{12}H_{22}O_{11}$ and hydrate formula is $C_{12}.11H_2O$. Lactose hydrolyzed by the lactase enzyme, that is a beta galactosidase (β- galactosidase term (lactase-phlorizin hydrolase) lactase) and the lactose is hydrolyzed to monosaccharide in the small intestine of body and localized to the tips of Villi.[8,26,30,41]

## LACTOSE

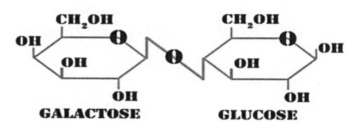

**FIGURE 10.1**   Structure of lactose.

Lactose intolerance also known as lactase deficiency is caused due to the deficiency of enzyme lactase in body.[8,10] Lactose intolerance is most common food intolerance and it is widely spread in the world. The 65% adults are affected with lactose intolerance in the world.[31] The humans with lactose intolerance loose the digestive capacity of lactose because of a genetically inadequate amount of lactase enzyme. Lactase deficiency is present in up to 100% of American Indians and Asians; up to 80% of blacks and Latinos; and 15% of Northern Europeans.[36]

Lactose intolerance is prevalent among infants and childrens with acute diarrheal disease. Lactose intolerance involves the immune system and causes injury to the intestinal surface. The 2–5% of infants are intolerant to cows milk till 1–3 months of age.[10,27–29] Lactose intolerance[42,43] depends on the quantity of consumed milk product lactose and the degree of lactase deficiency.[26] An amount of 50 g load of lactose in 1 L of milk causes problems like diarrhea, bloating, abdominal pain, and flatulence in majority of persons with lactose malabsorption. Lactose malabsorption is a physiologic problem that is attributed to lack of balance between the ingested lactose and hydrolyzed lactose. Unabsorbed lactose is substrate for intestinal mucosal bacteria, and they produce volatile fatty acids and gases ($CH_4$, $CO_2$, and $H_2$) leading to inflatulence.[26]

## 10.2.2  LACTASE DEFICIENCY

The absences or decline of intestinal lactase enzme[52] called lactase deficiency or hypolactasia. The lactase enzyme also called as lactose phlorizin hydrolase, β-galactosidase responsible for hydrolysis of lactose, the lactase enzyme have two sites, one of them hydrolyzing and other phlorizin (an aryl alpha glucoside). It is present on the apical surface of enterocyctes in the small intestine with highest expression in the mid jejunum and the c-terminal end protecting it. Lactose enzyme produced 220 kDa precursor peptides, which help in modification during post-transcription.[51] Lactase deficiency are divided into three parts (pathogenesis) as follows:

### 10.2.2.1  PRIMARY LACTASE DEFICIENCY

It is autosomal recessive and also known as adult hypolactasia, lactase nonpersistence[9] or hereditary lactase deficiency. It develops at various stages and approximately 70% of the world population suffers from primary lactase deficiency.[22] Lactase reduces by up to 65–90% in early childhood and keeps on declining throughout life. Evidence of lactase deficiency has been observed among: Hispanic poplation 50–75%, in Black and Ashkenazim people 60–80% and 20–25% of Hispanic, Asian and Black children younger than 4–5 years of age.[34]

### 10.2.2.2  SECONDARY LACTASE DEFICIENCY

Secondary lactase deficiency is a pathophysiologic condition. It is responsible for lactose intolerance. It is more common in children with acute gastroenteritits, damage of intestinal mucosa (acute viral or bacterial gastroenterits), uncontrolled coeliac disease or inflammatory bowel disease.[5,7]

### 10.2.2.3  CONGENITAL LACTASE DEFICIENCY

It is very little known about the molecular basics and is an extremely rare or single autosomal recessive disorder, where there is permanent absence of lactase activity.[25] This type of deficiency can be detected at the time of birth by feeding lactose to infant; and diarrhea will confirm the presence of this disorder.

### 10.2.3 SYMPTOMS OF LACTOSE INTOLERANCE

As discussed in the previous sections of this chapter, lactose intolerance is different from milk allergy; and it arises from lactose rather than milk proteins. It is also not like coeliac disease, an autoimmune disease caused by gluten. While travelling through small intestine, the lactose molecules, which due to some reason remain undigested, triggers the process of osmosis and consequently water is drawn out, which leads to the conditions of diarrhea and muscle cramps. Symptoms of lactose intolerance are: bloating, nausea, excessive flatulence, abdominal pain, and diarrhea.

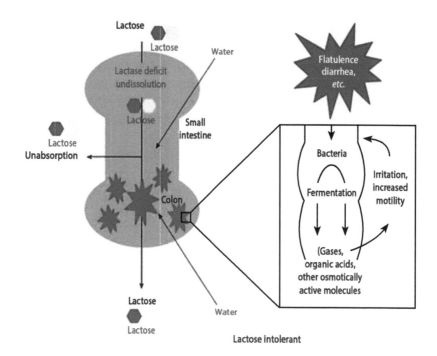

**FIGURE 10.2**  Mechanism of action in lactose intolerance.

Fermentation and metabolism of lactose by bacteria in the colon may also lead to such symptoms. During this process, gases, and some times short chain fatty acids are also produced (Fig. 10.2) usually during first couple of weeks from the birth and after this up to few months gastrointestinal symptoms begins to appear if feeding cows milk is not stopped. On ingestion of lactose-based foods, the symptoms of lactose intolerence begins to show within 2–4 h which can be confirmed by onset of vomiting, abdominal pain, diarrhea, nausea, bloating, flatus, and borborygmi. More common symptoms are: (1) abdominal pain, (2) diarrhea, and (3) flatulence (passing gas). Less common symptoms are: (1) abdominal bloating, (2) abdominal distention, and (3) nausea.

## 10.2.4   DIAGNOSIS OF LACTOSE INTOLERANCE

Physicians when suspect lactose intolerance, may conduct following diagonistic tests based on symptoms of the individual.

### a.   Lactose tolerance test

Blood glucose level in the body will rise just few hours after taking lactose-based food and drinks. Blood glucose concentration can be measured and if the otherwise case is found, it can be concluded that body is unable to digest lactose properly.

### b.   Hydrogen breath test

Hydrogen breath test requires the consumption of milk-based products with higher concentration of lactose, then the technician measures the level of hydrogen. The level of hydrogen in breath can be measured periodically. If the lactose remains undigested in the body, the fermentation in the lower intestine releases hydrogen and other gases, which are eventually exhaled. The exhaled hydrogen is measured in the breath that indicates that body is unable to digest and absorb lactose.

### c.   Stool acidity test

This is usually performed when infants and toddlers cannot undergo any another test. Lactic acid and other acids produced due to the fermentation of undigested lactose can be detected in the stool.

*d.   Intestinal biopsy test*

This test is useful for direct measurement of lactose through biopsy of the intestinal lining.

## 10.2.5   TREATMENT OF LACTOSE INTOLERANCE

At present, no effective treatment of lactose intolerance is available but we can avoid the discomfort of lactose intolerance by:

- Avoiding more milk and milk-based dairy products.
- Eating and drinking lactose reduced icecream and milk.
- Avoiding lactose containg foods and drinks, for example, salad cream, biscuits, chocolates, cakes, etc.
- To consume milk without any problem, add one spoon of sugar (white sugar) in the milk before the process of heating has begun. Once milk and sugar have been heated together, milk is drinkable or edible.

## 10.3   MILK ALLERGY

Often, people enjoy wide varieties of food with no problems. Certainly, for a minute percentage of people; however, specific food products or some ingredients of food may result in an adverse reaction ranging from rashes to severe allergic responses. Various adverse reactions to food products may be due to the food allergy. Food allergy is a specific type of intolerance that is due to a food or food ingredients that prompt and triggers the immune system. An allergen is responsible for activating various sets of reactions in the immune system which includes antibody synthesis as well.

Milk allergy is a form of food allergy in which various types of detrimental immune reactions to one or more than one components of milk or milk products from each and every animal (usually alpha S1-casein, a protein in cow's milk) takes place. Allergic reactions induced by milk result in anaphylaxis (refers to as potentially life-threatening condition). A person suffering from milk allergy can be reactive to one or more proteins within milk. The most frequently occurring type is alpha-S1-casein. *Bos domesticus* (cow) has about 30–35 g of proteins per liter in its milk. Either the activity of chymosin or milk acidification, which results in pH 4.6, enables the two portions to be generated from cow's milk proteins, such as:

- **Whey protein**: Globular proteins are the main components in whey. The major types are
- **β-lactoglobulin (β-Lg) and α-lactalbumin (α-Lac)**. Synthesis takes place in mammary gland, while other proteins, such as bovine serum albumin (BSA), immunoglobulins (Igs), lactoferrin (Lf), etc., are obtained from the blood source.
- **Casein**: Casein fraction (Cas) has four proteins in the coagulum, which are coded by the same chromosome. It includes: αS1-, αS2-, β-, and κcaseins. Casein proteins are highly phosphorylated with a loose tertiary and a hydrated structure. They can with stand and are not significantly affected by severe heat treatments.

At the time period of infancy and early childhood, milk allergy from cow is significantly most occurring type of food allergy. During infancy, the present immunoglobulin E (IgE) antibody specific for whey proteins and casein is liable for many detrimental responses to milk and milk products.[22,37,48] The β-Lg[24] mark the most important milk allergy protein because it is nowhere present in milk from human[22,24] and this allergen (β-Lg) gives the maximum estimate (66%) of severe oral threat in children having a milk allergy developed. However, it has been reported that casein protein has the highest estimate (68%) for adverse skin test reactivity.[13,14,49] The main features of various milk proteins are characterized in Table 10.1.

**TABLE 10.1** Main Characteristics of Major Milk Proteins.

| Protein (concentration %) | | Concentration estimated (g/L) | Molecular weight of protein (kDa) | No. of amino acids per molecule |
|---|---|---|---|---|
| Whey protein (20%) | β-Lg (10%) | 3–4 | 18.3 | 162 |
| | α-Lac (5%) | 1–1.5 | 14.2 | 123 |
| | Igs (3%) | 0.6–1.0 | 150 | – |
| | BSA (1%) | 0.1–0.4 | 66.3 | 582 |
| | Lf (traces) | 0.09 | 76.1 | 703 |
| Whole casein (80%) | α – S1-Cas (32%) | 12–15 | 23.6 | 199 |
| | α –S2-Cas (10%) | 3–4 | 25.2 | 207 |
| | κ-Cas (10%) | 3–4 | 19.0 | 169 |

## 10.3.1   TECHNIQUES AFFECTING COW'S MILK ALLERGENICITY

Milk from cow has been reported with high nutritive value and has wide range of bioactive properties; however, for some sensitized or immuno-compromised patients this can also act as a severe allergenic factor.[50] Wide range of techniques has been employed to decrease the allergenicity developed from cow's milk food products, such as:

- Lactic acid fermentation;
- Homogenization;
- Heat and enzymes treatment;
- Gamma irradiation; and
- Microwave and ultrasonic processing methods.

## 10.3.2   SYMPTOMS OF MILK ALLERGY

Often, children with milk allergy will show a slow reaction, which implies symptoms will develop over a time, perhaps within several hours to days, and these symptoms are comparably mild while certain people experience a life-threatening allergic reaction referred as anaphylaxis. Anaphylactic reaction sometimes occurs within just few minutes of exposure. It requires an urgent medical attention and it is clinically treated with epinephrine (EpiPen) in the form of a shot. Symptoms of milk allergy are as follows:

a.   *Respiratory*

- Nasal congestion or sneezing, runny nose
- Asthma
- Breathing difficulties
- Coughing
- Wheezing

b.   *Skin*

- Urticaria (hives)
- Swelling of the lips, tongue, mouth, face and/or throat
- Rashes or redness
- Eczema
- Itching (pruritus)

*c.* *Gastrointestinal*

- Diarrhea
- Abdominal cramps
- Vomiting
- Nausea
- Bloating
- Colic

*d.* *Systemic*

- Anaphylactic shock reaction (anaphylaxis)

## 10.3.3 DIAGNOSIS OF MILK ALLERGY

The milk allergy diagnostic tests are blood test and skin-prick test. Both tests detect the existence of IgE antibody, which is being generated, when body comes in contact with a substance to which it is particularly sensitive. These antibodies (immunoglobulins) prompt the discharge of various chemicals or signals that cause severe allergic symptoms.[12,21,23]

## 10.3.4 PROCESSING TECHNOLOGIES FOR TREATMENT OF MILK ALLERGENS

Allergen components of milk proteins[45–47] can be modified or removed to some extent by various processing techniques like enzyme hydrolysis, heat treatment, or pressure treatment etc. Due to lack of human studies, the effects of various treatments on the milk protein allergens determined by extensive *in vitro* studies are discussed here.

### 10.3.4.1 THERMAL PROCESSING

Thermal processing is one of the oldest and also among the most effective techniques for processing raw food. High temperature denatures the unwanted enzymes and inactivates the spoilage microbes thereby enhancing the shelf life of food. Proteins are also modified and become more digestible when food is cooked at high temperature. Heat also affects the allergenic and antigenic properties of proteins. However, pasteurization temperatures (whether it be flash pasteurization (95–97°C for 12 s.) or high temperature short time (HTST) (72°C for 15 s.) are not enough to reduce allergenicity of milk proteins.[11,35] However, boiling skimmed milk for 10 min with crossed radio-immuno electrophoresis (CRIE) showed that IgE-binding was decreased by 50 and 66% of the allergenic proteins α-Lac and α-Cas,[44] respectively, and it also eliminated β-Lg and serum albumin IgE-binding potencies completely. However, 2 and 5 min boiling time did not produce any significant alterations.[32,48] Thus, it can be concluded that thermal processing alone cannot be used for complete removal of allergenicity and it has to be applied in combination with other techniques like enzyme hydrolysis so that complete removal of allergenicity can be attained at normal pasteurization temperatures.

## 10.3.4.2   EFFECT OF ENZYMATIC HYDROLYSIS

Milk protein allergenicity[33] can be removed to a great extent by hydrolyzing with various proteolytic enzymes.[36] *In vitro* digestion[2] with duodenal fluid, elastase, and human trypsin followed by β-Lg and α-Lac casein showed the highest rate of degradation of milk protein allergens.[17,18] Stability with more than 60 min has been reported for β-Lg against peptic hydrolysis (pH 1.2), while casein and serum albumin got completely hydrolyzed after 2 min and 30 s, respectively.

## 10.3.4.3   EFFECT OF GAMMA IRRADIATION

Irradiation modifies the structure of food proteins and studies have shown that ionizing radiations can change allergenicity by modifying the antibody binding epitopes in food allergens.[4,15,16,40] Gamma irradiation[6] reduces the amount of intact allergens in a solution depending on the dose. The sensitivity of α-lactose and β-Lg is higher than the other proteins at same dose of irradiation. Thus, it can be concluded that gamma-irradiation is a promising tool to curb milk allergies.

## 10.3.4.4   EFFECT OF ULTRAFILTRATION

Ultrafiltration followed by enzymatic hydrolysis gave good results in terms of reduced allergen content in milk samples.[15] Ultrafiltration followed by heat processing and proteolysis[1] is an efficient method to remove milk allergens. Therefore, ultrafiltration is not an independent solution to the problem but it compliments other methods and enhances their efficiency.

## 10.3.4.5   EFFECT OF PRESSURE

Homogenization is the technique in which high pressure is used to crush the fat globules in order to reduce their size and facilitate their uniform distribution throughout the milk. It helps to increase the shelf life as well as the palatability of the milk. Anaphylactic reaction was triggered at a dose of 340 µg homogenized and pasteurized milk, in a murine model.[33] Investigators also found that higher fat content resulted in more anaphylactic reactions at same dose of homogenized milk. However, there is no evidence of any effect of homogenization on the milk protein allergens[19] but allergic children digest less homogenized milk more effectively.[26] On the other hand, some studies suggest that there is no significant difference between the groups treated with homogenized and non-homogenized milk.

## 10.3.4.6   EFFECT OF MICROWAVE AND ULTRASONIC TECHNOLOGIES

For treatment of milk allergens, ultrasonic and microwave technologies have proved to be more effective than low-temperature long-time (LTLT) pasteurization[49] (90°C, 15 min). The α-Lac and β-Lg immunoreactivity were reduced to 0.88 and 6.42%, respectively, by ultrasonic treatment at 50°C for 1 h, whereas the α-Lac and β-Lg immunoreactivity were reduced to 1.37 and 12.86%, respectively by microwaving at 98°C for 2 min. The indirect and competitive enzyme-linked immunosorbent assay (ELISA) was used to detect immunoreactivity.

## KEYWORDS

- allergencity
- allergens
- enzymatic hydrolysis
- food allergy
- immunoglobulin E (IgE)
- lactose intolerance

## REFERENCES

1. Asselin, J.; Hebert, J.; Amiot, J. Effects of in Vitro Proteolysis on the Allergenicity of Major Whey Proteins. *J. Food Sci.* **1989,** *54* (4), 1037–1039.
2. Astwood, J. D.; Leach, J. N.; Fuchs, R. L. Stability of Food Allergens to Digestion in Vitro. *Nat. Biotechnol.* **1996,** *14 (*10), 1269–1273.
3. Bernard, H.; Creminon, C.; Negroni, L.; Peltre, G.; Wal, J. M. IgE Cross-Reactivity with Caseins from Different Species in Humans Allergic to Cow's Milk. *Food Agric. Immunol.* **1999,** *11* (1), 101–111.
4. Beslr, M.; Steinhart, H.; Paschke, A. Stability of Food Allergens and Allergenicity of Processed Foods. *J. Chromatogr B. Biomed. Sci. Appl.* **2001,** *756* (1–2), 207–228.
5. Bhatnagar, S.; Bhan, M. K.; Singh, K. D. Efficiency of Milk Based Diets in Persistent Diarrhea: A Randomized, Controlled Trial. *Pediatrics.* **1996,** *98,* 1122–1126.
6. Byunm, W.; Lee, J. W.; Yookh, J. C.; Kim, H. Y. Application of Gamma Irradiation for Inhibition of Food Allergy. *Radiat Phys. Chem.* **2002,** *63* (3–6), 369–370.
7. Cox, T. M. Disaccharide Deficiency. In *Oxford Textbook of Medicine,* 4th ed.; Warrell, D. A., Cox, T. M., Firth, J. D., Benz, E. J. Eds.; Oxford University Press: New York, NY, 2003; pp 593–597.
8. Dahlqvist, A.; Hammond, J. B. Intestinal Lactase Deficiency and Lactose Intolarance in Adults: Preliminary Report. *Gastroenterology.* **1963,** *45,* 488–491.
9. Dallas, M. Swallow Genetics of Lactose Persistence and Lactose Intolarance. *Annu. Rev. Genet.* **2003,** *37,* 197–219.
10. Edmond, H. M. M.; Richard, J. Lactose Intolarance and Lactase Deficiency in Children. *Curr. Opin. Pediatr.* **1994,** *6,* 562–567.
11. Fiocchi, A.; Bouygue, R.; Sarratud, T.; Terracciano, L.; Martelli, A.; Restani, P. Clinical Tolerance of Processed Foods. *Ann. Allergy Asthma Immunol.* **2004,** *93* (Suppl. 3), S38–S46.
12. Gjesing, B.; Osterballe, O.; Schwartz, B.; Wahn, U.; Lowenstein, H. Allergen Specific IgE Antibodies Against Antigenic Components in Cow Milk and Milk Substitutes. *Allergy.* **1986,** *41* (1), 51–56.

13. Goldman, A. S.; Anderson, D. W., Sellers, W. A.; Saperstein, S.; Kniker, W. T.; Halpern, S. R. Milk Allergy, Part I-Oral Challenge with Milk and Isolated Milk Proteins in Allergic Children. *Pediatrics.* **1963,** *32,* 425–443.

14. Goldman, A. S.; Sellars, W. A.; Halpern, S. R.; Anderson, D. W.; Furlow, T. E.; Johnson, C. H. Milk Allergy, Part II-Skin Testing of Allergic and Normal Children with Purified Milk Proteins. *Pediatrics.* **1963,** *32,* 572–579.

15. Gortler, I.; Urbanek, R.; Forster, J. Characterization of Antigens and Allergens in Hypo-Allergenic Infant Formulae. *Eur. J. Pediatr.* **1995,** *154* (4), 289–294.

16. Hochwallner, H.; Schulmeister, U.; Spitzauer, S.; Valentaa, R. Cow's Milk Allergy: From Allergens to New Forms of Diagnosis, Therapy and Prevention. *Methods.* **2014,** *66* (1), 22–33.

17. Jakobsson, I.; Borulf, S.; Lindberg, T.; Benediktsson, B. Partial Hydrolysis of Cow's Milk Proteins by Human Trypsins and Elastases in Vitro. *J. Pediatr. Gastroenterol Nutr.* **1983,** *2* (4), 613–616.

18. Jakobsson, I.; Lindberg, T.; Benediktsson, B. In Vitro Digestion of Cow's Milk Proteins by Duodenal Juice from Infants with Various Gastrointestinal Disorders. *J. Pediatr. Gastroenterol Nutr.* **1982,** *1* (2), 183–191.

19. Jost, R. Physicochemical Treatment of Food Allergens: Application to Cow's Milk Proteins. *Nestlé Nutr. Workshop Ser.* **1988,** *17,* 187–197.

20. Kilara, A.; Panyam, D. Peptides from Milk Proteins and their Properties. *Crit. Rev. Food Sci. Nutr.* **2003,** *43* (6), 607–633.

21. Kohno, Y.; Honma, K.; Saito, K. Preferential Recognition of Primary Protein Structures of Alpha-Casein by IgG and IgE Antibodies of Patients with Milk Allergy. *Ann. Allergy.* **1994,** *73* (5), 419–422.

22. Kretchmer, N. Lactose and Lactase. *Sci. Amer.* **1971,** *227,* 70–78.

23. Kuitunen, M.; Savilahti, E. Mucosal IgA, Mucosal Cow's Milk Antibodies, Serum Cow's Milk Antibodies and Gastrointestinal Permeability in Infants. *Pediatr. Allergy Immunol.* **1995,** *6,* 30–35.

24. Kuitunen, M.; Savilahti, E.; Sarnesto, A. Human α-lactoalbumin and Bovine β-lactoglobulin Absorption in Infants. *Allergy.* **1994,** *49,* 354–360.

25. Lee, J. W.; Yook, H. S.; Lee, K. H.; Kim, J. H.; Byun, M. W. Conformational Changes of Myosin by Gamma Irradiation. *Radiat Phys. Chem.* **2000,** *58* (3), 271–277.

26. Lomer, M. C. E.; Parkes, G. C.; Sanderson, J. D. Review Article: Lactose Intolarance in Clinical Practice—Myths and Realities. *Aliment. Pharm. Therap.* **2008,** *27* (2), 91–103.

27. Mattar, R.; Ferraz, D.; Mazo, D.; Carriho, F. J. Lactose Intolarance: Diagnosis, Genetics and Clinic Factor. *Clin. Exp. Gastroenterol.* **2012,** *5,* 113–121.

28. McCracken, R. D. Lactose Deficiency; an Example of Dietary Evolution. *Curr. Anthropol.* **1971,** *12,* 479–517.

29. Melvin, B. Lactose Intolarance in Infants, Children, and Adolescents. *Pediatrics.* **2006,** *62* (2), 240.

30. Michalski, M. C. On theSsupposed Influence of Milk Homogenization on the Risk of CVD, Diabetes and Allergy. *Br. J. Nutr.* **2006,** *97,* 598–610.

31. National Dairy Council–India. An Interpretive Review of Recent Nutrition Research: Prospective on Milk Intolarance. *Dairy Counc. Dig.* **1978,** *49* (6), 120–128.

32. Newcomar, A. D.; Mcgill, D. B. Lactose Tolerance Tests in Adults with Normal Lactase Activity. *Gastroenterology.* **1996,** *50,* 340–346.

33. Norgaard, A.; Bernard, H. Allergenicity of Individual Cow Milk Proteins in DBPCFC-Positive Milk Allergic Adults. *J. Allergy Clin. Immunol.* **1996,** *97,* 2371.

34. Paige, D. M.; Bayless, T. M.; Mellitis, E. D.; Davis, L. Lactose Malabsorption in Pre School Black Children Amj. Clin. *Nutr.* **1977,** *30,* 1018–1022.

35. Poulsen, O. M.; Hau, J.; Kollerup, J. Effect of Homogenization and Pasteurization on the Allergenicity of Bovine Milk Analyzed by a Murine Anaphylactic Shock Model. *Clin. Allergy.* **1987,** *17,* 449–458.

36. Rana, S.; Malik, A. Breath Tests and Irritable Bowel Syndrome. *World J. Gastroenterol.* **2014,** *20* (24), 7587–7601.

37. Sahi, T. Genetics & Epidemology of Adult Type Hypolactasia. *Scand J. Gastroenterol Suppl.* **1994,** *202,* 7–20.

38. Sampson, H. A. IgE-Mediated Food Intolerance. *J. Allergy Clin. Immunol.* **1988,** *81* (3), 495–504.

39. Schmidl, M. K.; Taylor, S. L.; Nordlee, J. A. Use of Hydrolyzate-Based Products in Special Medical Diets. *Food Technol. (USA).* **1994,** *48* (10), 77–85.

40. Schütte, L.; Paschke, A. *Reducing Allergens in Milk and Milk Products;* Woodhead Publishing Ltd.: Sawston, Cambridge, 2004; Vol. 1, pp 159–177. doi:10.1533/9781845692278.2.159

41. Shaukat, A.; Levitt, M. D.; Taylor, B. C.; MacDonald, R.; Shamliyan, T. A.; Kane, R. L.; Wilt, T. J. Effective Management Strategies for Lactose Intolarance. *Evid. Rep. Technol. Assess.* **2010,** *152* (192), 1845–1850.

42. Suarez, F. L.; Savaiano, D. A.; Levitt, M. D. A Comparison of Symptoms after the Consumption of Milk or Lactose Hydrolyzed Milk by People with Self Reported Severe Lactose Intolarance. *N. Engl. J. Med.* **1995,** *333* (1), 1–4.

43. Swagerty, D. L.; Walling, A. D.; Klein, R.M. Lactose Intolarance. *Am. Fam. Physician.* **2002,** *52,* 797–803.

44. Tada, Y.; Nakase, M.; Adachi, T. Reduction of 14–16 kDa Allergenic Proteins in Transgenic Rice Plants by Antisense Gene. *FEBS Lett.* **1994,** *391* (3), 341–345.

45. Wal, J. M. Cow's Milk Allergens. *Allergy.* **1988,** *53* (11), 1013–1022.

46. Wal, J. M. Cow's Milk Proteins/Allergens. *Ann Allergy Asthma Immunol.* **2002,** *89* (6), 3–10.

47. Wal, J. M. Structure and Function of Milk Allergens. *Allergy.* **2001,** *56* (67), 35–38.

48. Wang, Y.; Harvey, C. B.; Hollox, E J. The Genetically Programmed Down-Regulation of Lactose in Children. *Gastroenterology.* **1998,** *114,* 1230–1236.

49. Werfel, S. J.; Cooke, S. K.; Sampson, H. A. Clinical Reactivity to Beef in Children Allergic to Cow's Milk. *J. Allergy. Clin. Immunol.* **1997,** *99* (3), 293–300.

50. Wroblewska, B.; Jedrychowski, L. Changes of Immunoreactive Properties of Cow Milk Proteins as a Result of Technological Processing. *Biotechnologia.* **2001,** *3,* 189–201.

51. Wroblewska, B.; Jedrychowski, L. Influence of Technological and Biotechnologi-Cal Processes on the Immunoreactivity of Cow Milk Whey Proteins. *Pol. J. Food Nutr. Sci.* **2002,** *11* (2), 156–159.

52. Zecca, L.; Mesonero, J. E.; Stutz, A.; Poirée, J.-C.; Giudicelli, J.; Cursio, R.; Gloor, S. M.; Semenza, G. Intestinal Lactase-Phlorizin Hydrolase (LPH): The Two Catalytic Sites; the Role of the Pancreas in Pro-LPH Maturation. *FEBS Lett.* **1998,** *435,* 225–228.

# PART IV

# Management Systems and Hurdles in the Dairy Industry

# CHAPTER 11

# DAY LIGHTING IN DAIRY FARMS

BHAVESH B. CHAVHAN[1*], A. K. AGRAWAL[2], C. SAHU[2],
S. S. CHOPDE[3], and ADARSH M. KALLA[4]

[1]*Department of Dairy Engineering, College of Dairy Technology,
Maharashtra Animal and Fisheries Sciences University, Udgir
413517, Nagpur, India*

[2]*Department of Dairy Engineering, College of Dairy Science and
Food Technology, Chhattisgarh Kamdhenu Vishwavidyalaya
(CGKV), Raipur 492012, India*

[3]*Department of Dairy Engineering, College of Dairy Technology
at Udgir, COVAS Campus, Kavalkhed Road, Udgir 413517,
Maharashtra, India*

[4]*Department of Dairy Engineering, Dairy Science College,
Karnataka Veterinary Animal and Fisheries Sciences University,
Mahagaon Cross, Kalaburagi 585316, Karnataka, India*

*\*Corresponding author. E-mail. bhaveshchavhan@gmail.com;
bhaveshchavhan90@gmail.com*

## CONTENTS

## ABSTRACT

Photoperiod manipulation, also known as long-day lighting (LDL), is the management practice of using a designed lighting system to artificially extend the duration of light that a lactating cow is exposed to light with the goal of increasing milk production. Photoperiod manipulation has gained much wide span industry interest.[4] Appropriate lighting can improve productivity and safety on a dairy farm. On average, lighting represents 17% of total dairy farm electrical energy use. Optimal lighting conditions may increase milk productivity and conserve energy consumption. Factors that contribute to increased milk production include: the type of light, the amount of light provided per watt, the temperature of the work area, the height of the ceilings, and the length of the lighting period. By making these changes, along with other improvements, the dairy operation and lighting system will be more energy efficient. This can lead to improved farm productivity and increased revenue, while lowering energy costs.

## 11.1   INTRODUCTION

Photoperiod manipulation, also known as long-day lighting (LDL), is the management practice of using a designed lighting system to artificially extend the duration of light that a lactating cow is exposed to light with the goal of increasing milk production. Photoperiod manipulation was first pioneered in the late 1970s and gained much wide span industry interest in the late 1990s. Supplementing lactating cows with 16–18 h of continuous light at an intensity of 15–20 foot candles (fcs) has shown to increase milk production from 5 to 16% compared to cows exposed to less than 13.5 h of light in the research trials. Research has also shown that the balance of the 24 h period needs to be dark in order to achieve a favorable response.

Appropriate lighting can improve productivity and safety on a dairy farm. On average, lighting represents 17% of total dairy farm electrical energy use. Optimal lighting conditions may increase milk productivity and conserve energy consumption. Factors that contribute to increased milk production include: the type of light, the amount of light provided per watt, the temperature of the work area, the height of the ceilings, and the length of the lighting period. By making these changes along with other improvements, the dairy operation lighting system will be more energy efficient.

This can lead to improved farm productivity and increased revenue, while lowering energy costs. Dairy producers are constantly searching out new management techniques to improve production efficiency and cash flow.

Brightness is a measure of the amount of light striking a surface. It is measured in units of fcs. One fc is defined as 1 lumen of light falling on 1 square foot. The metric equivalent of one fc is a lux (lumen per square meter). One *fc* = 10.76 lux.

Photoperiod management has received interest as a cost effective method to increase production in lactating cows. That is because in cows exposed to long days (16–18 h of light (180 lux)) and a 6–8 h period of darkness; daily milk production increases an average of 2 L/cow, relative to those on natural photoperiods.[6] The 15–20 fcs has been shown to increase milk production from 5 to 16% compared to cows exposed to less than 13.5 h of light.[3,9,12,13,15,16]

The purpose of this chapter is to: review the evidence for a response of lactating and dry cows to photoperiod; describe the physiologic basis for those responses; and discuss the energy efficient lighting use in farm so to improved farm productivity and increase revenue, while lowering energy costs.

## 11.2   EFFECTS OF PHOTOPERIOD IN LACTATING COWS

Effects of photoperiod in lactating cows have been studied by researchers at Michigan State University (MSU).[12] In cows, a long day actually reduces the duration of elevated melatonin;[6] and a long day pattern was related to higher secretion of the hormone insulin-like growth factor-I (IGF-I),[7] which is thought to increase milk yield. The bovine somatotropin (bST) increases milk yield and stimulates IGF-I release.[3] The combination of long days and bST was able to increase production 7.7 L/d.[10]

Dahl in http://www.wcds.ca/proc/2003/Manuscripts/Chapter%2027% 20Dahl.pdf indicates that *"combining long days with other management techniques may improve performance. On one farm with 2000 cows, long days increased production relative to those cows that remained on natural lighting."*[5–8] Dahl[5–8] also indicates that *"Lactating cows should be under long day photoperiod of 16 to 18 hours of light to increase milk production. In late pregnancy expose cows to short day photoperiod of less than 10 hours of light to maximize production and improve health status in the transition period."*

Recent studies suggest that appropriate photoperiod treatment of the dry cow can markedly enhance milk yield in the subsequent lactation.[1,5–8,11,14]

These data suggest that short days are associated with greater resistance to pathogenic insult during an immune-compromised period in the production cycle. Shifts in secretion of and sensitivity to the hormone prolactin (PRL) may explain the effects of short day photoperiod during the dry period. It has been observed that long days increase whereas short days decrease PRL secretion.[2] However, the lower PRL concentration of cows on short days is associated with higher amounts of PRL-receptor expression, and likely sensitivity to that hormone.[2] Because PRL is critical to the process of mammary cell activation that occurs at parturition, and PRL has immunostimulatory effects, it is speculated that the shifts in sensitivity that accompany short day treatment are producing the changes observed in production and mammary health.

## 11.3   ENERGY EFFICIENT LIGHT FOR DAIRY FARMS

The common theme behind the use of all these sources is the basic need for supplemental light to provide the visual acuity to perform required functions accurately, efficiently, and safely. As the continuing trend toward larger dairies operating around the clock continues the necessity of efficient, well-designed and maintained lighting systems becomes even more crucial to successful operation of the farm. The available energy conservation options for improving lighting efficiency and efficacy on the farm are enormous. New and improved lighting technology is being developed continually. The choices available range from simple lamp replacements to installing new high-efficiency lighting systems with programmable logic controllers and other computer-based control systems. An integral step to improving lighting on dairy farms is the performance of a specific lighting design for that area or facility. Clearly using an efficient light source in a poor design does not provide optimal lighting. Thus design should satisfy established criteria for light level, color rendering, efficacy, selection of fixtures suitable for the ambient environment, controls and proper wiring, and circuit protection. A recent development in the application of lighting technology on dairy farms may involve photoperiod manipulation, or LDL, of dairy cows to increase milk production. This management practice uses an increased light intensity over a defined time interval to stimulate increased milk production. An efficient lighting design and control system must be implemented to obtain the benefits of LDL. The type of luminaire that can used to provide lighting includes metal halide (MH), high-pressure sodium (HPS), and fluorescent. These three light sources will create a milk yield response.[8] There may be personal preferences with regard to MH and HPS luminaires. HPS lights deliver more lumens per Watt than MH. This is obviously an economic advantage because fewer luminaires are needed to obtain the same average light level (number of lumens measured at the work plane) and lower connected load (kW). However, achieving "uniform" lighting is easier to provide when using more, lower wattage fixtures. Generally, HPS luminaires also have a higher average lamp life. However, MH have a higher color rendition index (CRI), a measure of the luminaire's output light color. MH luminaires have a more white output light color while HPS luminaires have a yellowish/orange output color. Lighting system designers should consult with the dairy producer and determine their preference for luminaire output light color.

Fluorescent lights have a lamp life similar to MH. However, fluorescent lamps have a lower lamp lumen depreciation factor meaning that they stay brighter longer. In fact, mean light output for fluorescent lamps can be 95% of the initial light output while MH may be 65%. Mean light output is defined as the light output after 40% of average life. The rated average life for fluorescent and MH lights is determined when 50% of the installed lights are still operating.

Finally, LED lights can provide high energy efficiency with a reported 100,000 h operating life. This is significantly longer than the reported 20,000 h operating life of fluorescent and HID fixtures. Moreover, LED lights are expected to have lower maintenance costs, contain no mercury, and provide instantaneous reliable light. However, LED fixtures are expensive compared to the other fixtures. These unique attributes can make it confusing to select fixtures best suited for dairy operations. When considering implementing LDPP, LED fixtures may provide an edge.

Lighting performance is often measured based on lumens/watt. This can be misleading for dairy producers because lumens represent effective light for the human eye. Dairy cows perceive light differently than humans, meaning a light fixture can provide ample lumens/watt, but may not provide light in the appropriate spectrum to stimulate milk yield. For instance, high pressure sodium fixtures provide high lumens/watt, however, light output from these fixtures is biased toward longer wavelengths that cows cannot perceive. Fluorescent fixtures provide ample effective light for the cow. However under cold conditions, light output of fluorescent fixtures can decrease by more than 40%. Cold conditions are typical in most barns during winter months, precisely when supplemental lighting from light fixtures is required. LED fixtures can provide light in the same spectrum as sunlight and are more reliable under cold conditions. These two considerations suggest LED fixtures may be best suited for implementing LDPP. However, this scenario needs to be investigated under barn conditions.

## 11.4   LIGHTING DESIGN

Lighting design is important as it will determine the performance of the lighting for the life of this system. Therefore, it is worth getting some professional advice to get the adequate lighting. The preceding sections have covered the issues of lighting level, light uniformity, shadows, color, and control—all these parameters must be considered in the design process. Following considerations must also be examined:

- Consider the way lighting is used on an everyday basis. Where is the most suitable area for the switches to be located? Is it possible to get different lighting levels by simply grouping and switching the lights on/off in clusters?
- Are the lights in a position where they can be easily cleaned and where the bulbs can be safely changed?
- Consider reflectivity off roofs and walls. Coloring surfaces white or a light color can increase the lighting level dramatically.
- Fittings in most cases will have to be water and dust proof. Make sure the ones you choose are up to standard.

## 11.5   LIGHTING REQUIREMENTS ON DAIRY FARMS

Table 11.1 gives some guidance on lighting levels and desirable light properties for different areas of a dairy enterprise. There are no definitive standards on this topic; therefore, table is based on data derived from practical experience and from similar practical references.

**TABLE 11.1** Lighting Requirements and Desirable Light Properties for Different Sections of a Dairy Enterprise.

| Application | Lux level required | Color rendering | Uniformity | Control | Comments |
|---|---|---|---|---|---|
| Cubicle and feeding area | 170–200 lux for photo period yield effect, 50 lux for general | Low to medium | Medium | Timed, with light level sensing. Fluorescents can use light level driven dimming | High pressure sodium, metal halide lights, or multiple fluorescent fittings |
| Milking area | 500 lux for pit | Good | Very good | Timed with manual override | Fluorescent lights will punch light through the mass of pipes and fittings and give even shadow-less light |
| Collection yard | 50 lux | Low to medium | Medium | Timed with manual override | High pressure sodium or metal halide lights |
| Bulk tank area | 200 lux | Good | Medium | Proximity | Fluorescent lights are most commonly used |
| Outside areas | 20 lux | Low to medium | Low | Timed/light level | High pressure sodium or metal halide lights are the best compromise between cost and performance |
| Office | 300–500 lux | Good | Good | Proximity | Fluorescent lights are most commonly used |

## KEYWORDS

- dairy farm
- day lighting
- electrical energy
- energy cost
- energy efficient
- foot candles

## REFERENCES

1. Aharoni, Y.; Brosh, A.; Ezra, E. Short Communication: Prepartum Photoperiod Effect on Milk Yield and Composition in Dairy Cows. *J. Dairy Sci.* **2000,** *83,* 2779–2781.
2. Auchtang, T. L.; Pollard, B. C.; Kendall, P. E. Prolactin Receptor Expression Responds to Photoperiod Similarly in Multiple Tissues in Dairy Cattle. *J. Anim. Sci.* **2002,** *80* (1), 9.
3. Bauman, D. E. Bovine Somatotropin and Lactation: From Basic Science to Commercial Application. *Domest. Anim. Endocrinol.* **1999,** *17,* 101–116.
4. Bilodeau, P. P.; Petiticlerc, D.; Pierre, N.; Laurent, G. J. Effects of Photoperiod and Pair-feeding on Lactation of Cows Fed Corn or Barley Grain in Total Mixed Rations. *J. Dairy Sci.* **1999,** *72,* 2999.
5. Dahl, G. E.; Petitclerc, D. Management of Photoperiod in the Dairy Herd for Improved Production and Health. *J. Anim. Sci.* **2003,** *80* (2), 110–113.
6. Dahl, G. E.; Buchanan, B. A.; Tucker, H. A. Photoperiodic Effects on Dairy Cattle – A Review. *J. Dairy Sci.* **2000,** *83,* 885.
7. Dahl, G. E.; Elsasser, T. H. Effects of Long Day Photoperiod on Milk Yield and Circulating Insulin-like Growth Factor – Part I. *J. Dairy Sci.* **1997,** *80,* 2784–2789.
8. Dahl, G. E. In *Photoperiod Control Improves Production and Profit of Dairy Cow,* Proceedings of the 5th Western Dairy Management Conference, April 4–6, DeLaval: Las Vegas, NV; 2001; pp 27–30.
9. Marcek, J. M.; Swanson, L. V. Effect of Photoperiod on Milk Production and Prolactin of Holstein Dairy Cows. *J. Dairy Sci.* **1984,** *67,* 2380.
10. Miller, A. R. E.; Stanisieaski, E. P.; Erdman, R. A.; Douglass, L.W. Effects of Long Day Photoperiod and Bovine Somatic (Trobest) on Milk Yield in Cows. *J. Dairy Sci.* **1999,** *82,* 1716.
11. Miller, A. R. E.; Erdman, R. A.; Douglass, L.W.; Dahl, G. E. Effect of Photoperiodic Manipulation Dairying the Dry Period of Dairy Cows. *J. Dairy Sci.* **2000,** *83*(May), 962–967.

12. Peters, R. R.; Chapin, L. T.; Emery, R. S.; Tucker, H. A. Milk Yield, Feed Intake, Prolactin, Growth Hormone, and Glucocorticoid Response of Cows to Supplemental Light. *J. Dairy Sci.* **1981,** *64,* 1671.
13. Peters, R. R.; Chapin, L. T.; Leining, K. B.; Tucker, H. A. Supplemental Lighting Stimulates Growth and Lactation in Cattle. *Science.* **1978,** *199,* 911.
14. Petitclerc, D.; Vinet, C.; Roy, G.; Lacasse, P. 1998. Prepartum Photoperiod and Melatonin Feeding on Milk Production and Prolactin Concentrations of Dairy Heifers and Cows. *J. Dairy Sci.* **1998,** *81*(1), 110–112.
15. Phillips, C. J. C.; Scholfield, S. A. The Effects of Supplemental Light on the Production and Behavior of Dairy Cows. *Anim. Prod.* **1989,** *48,* 293.
16. Stanisiewski, E. P.; Mellenberger, R. W.; Anderson, C. R.; Tucker, H. A. Effect of Photoperiod on Milk Yield and Milk Fat in Commercial Dairy Herds. *J. Dairy Sci.* **1985,** *68,* 1134.

**CHAPTER 12**

# REFRIGERATION PRINCIPLES AND APPLICATIONS IN THE DAIRY INDUSTRY

A. G. BHADANIA*

*Department of Dairy Engineering, Sheth M. C. College of Dairy Science, Anand Agricultural University, Anand 388110, India*

*E-mail: bhadania@gmail.com

## CONTENTS

## ABSTRACT

Refrigeration is the most important utility required for dairy plants for low temperature storage of different food and dairy products. It is also very essential for cold chain of handling of milk and other perishable food products. Vapor compression refrigeration system using ammonia as refrigerant is widely used in India for industrial refrigeration, air conditioning, and cold storages. In dairy processing plants, ice-bank system of refrigeration is used for chilling and processing of milk, while direct expansion air chillers are employed for cold storages. Multi-compression and evaporators systems are being used for the storage of icecream and many other frozen products. The refrigeration system uses electrical energy for operation of compressor and other auxiliary components of the system. Performance evaluation and efficient operational management of refrigeration plants are very important as electrical power consumption of refrigeration plant alone is about 50–60 % of total electrical consumption of the dairy and food plants, depending on the nature of processing operations. Vapor absorption refrigeration system can be used in dairy plants using waste thermal energy or using cogeneration system.

This chapter covers the working of vapor compression system, factors affecting performance of the refrigeration system, system components including ice-bank and ice silo, controls, maintenance, etc. The use of refrigerant properties, tables, and charts for thermodynamic analysis of vapor compression refrigeration system is also described. The calculations for estimating refrigeration plant capacity for bulk milk cooler, chilling center, cold storage, and dairy plant are also presented. The most commonly used refrigerants for vapor compression refrigeration systems and safety aspects of ammonia as refrigerant are also presented.

## 12.1   INTRODUCTION

The American Society of Heating, Air conditioning, and Refrigerating Engineers (ASHARE) defines *"refrigeration is the science of providing and maintaining temperatures below the ambienttemperatures."*[4] In other words; it is the process of cooling a substance below the initial temperature of the substance. Air conditioning refers to the control of environmental conditions of air, depending on the use of air conditioning.[2,3,19-21,24,29] Air conditioning signifies the control of an atmospheric environment either for humans, or to carry out industrial or scientific processes efficiently. Environmental air conditioning in terms of temperature, humidity, and purity of air results in greater comfort to occupants when applied to public places, offices, and factories.

Vapor compression refrigeration system using ammonia as refrigerant is widely used in India for industrial refrigeration, air conditioning[6,11,20,21,24,29] and cold storages. In dairy processing plants, ice-bank system of refrigeration is used for chilling and processing of milk, while direct expansion air chillers are being employed for cold storages,[9,10,23] Refrigeration is also very essential for cold chain of handling of many agricultural produce, especially fruits and vegetables. Single stage compression systems and two-stage vapor compression systems are used in dairy and food plants. Two-stage vapor compression system is used when compression ratio is higher (> 7) especially for frozen storage requirements of icecream and other food products. The refrigeration system uses electrical energy for operation of compressor and other auxiliary components of the system.[30]

This chapter covers working of vapor compression system, factors affecting performance of the refrigeration system, system components including ice-bank and ice silo, controls, maintenance, etc. The use of refrigerant properties, tables, and charts for thermodynamic analysis of vapor compression refrigeration system is also described. The calculations for estimating refrigeration plant capacity for bulk milk cooler,[1] chilling center, cold storage, and dairy plant are also presented. This chapter includes the discussion on most commonly used refrigerants for vapor compression refrigeration systems and safety aspects of ammonia as a refrigerant.

## 12.2  IMPORTANCE OF REFRIGERATION IN THE DAIRY INDUSTRY

Refrigeration is a basic requirement for the processing and storage of milk and milk products, as majority of dairy products are perishable in nature. The importance of refrigeration is indicated below.

a. Chilling of milk at producers' level by employing bulk milk coolers and at milk chilling centers is the first requirement in dairy industry. Immediate cooling of milk to about 2–3 °C is very important to reduce the multiplications of microorganisms and to get low bacterial count in milk and milk products.

b.  Processing of milk using either batch pasteurizer or HTST plant requires chilled water or any other cooling medium for cooling of milk.

c.  Manufacture of many products requires refrigeration. For example, butter, icecream, etc.

d.  Storage of milk and milk products requires maintaining low temperature in the cold storages depending on the type of product to be stored. For example, milk is stored at around 3–4°C while icecream is stored at−30°C temperature.

e.  Transportation of many products requires refrigerated vehicles to maintain the quality of products.

f.  Low temperature storage is required for distribution of products as well as at the consumers' level.

## 12.3   VAPOR COMPRESSION REFRIGERATION SYSTEM

The vapor compression refrigeration (VCR) system using ammonia as refrigerant is widely used in dairy plants. The system consists of compressor, condenser, receiver, expansion valve, evaporator, and automatic controls.[4,12] The liquid refrigerant is throttled from condensing pressure to evaporating pressure by the expansion valve and expanded ammonia evaporates in the evaporator by absorbing the heat from the space.[17,25] The vapor produced in the evaporator is pumped by the compressor and its pressure is raised to a level so as to condense it at the temperature of the cooling medium. The condensed refrigerant goes to the receiver and it is again supplied to the evaporator through the expansion valve. The liquid refrigerant absorbs the heat from a zone of low pressure (evaporator) by means of its evaporation (Figs. 12.1 and 12.2). The heat is dissipated in a zone of higher pressure (condenser) by means of condensation.[5] The refrigerants (like ammonia, R-22[1], R-134a, etc.) absorb the heat at the evaporator through evaporation.[1,4,12]

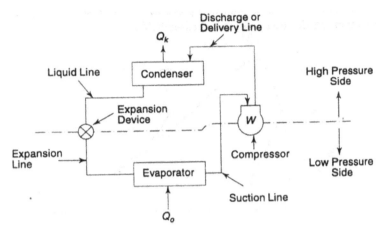

**FIGURE 12.1**   Flow chart for a typical refrigeration system.

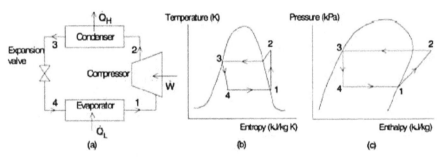

**FIGURE 12.2**   Basic vapor pressure compression refrigeration system (a); T–s diagram (b); Log P-h diagram (c).

## 12.4   UNITS OF REFRIGERATION

The capacity of refrigeration plant required in any dairy/food plant can be estimated based on the cooling load requirement of the plant. The capacity of refrigeration system is expressed as ton of refrigeration (TR). A ton of refrigeration is defined as the quantity of heat to be removed in order to form one ton (2000 lb) of ice at 0°C in 24 h, from liquid water at 0°C. This is equivalent to 12,600 kJ/h or 210 kJ/min or 3.5 kJ/s (3.5 kW).

$$\text{One } TR = 12,600 \text{ kJ/h or } 210 \text{ kJ/min or } 3.5 \text{ kW} \qquad (12.1)$$

## 12.5   COMPONENTS OF A VAPOR COMPRESSION REFRIGERATION SYSTEM

The refrigeration system is a closed system consisting of different basic components to accomplish the various processes required for the refrigeration. The basic components of VCR system are: compressor, condenser, receiver, expansion valve, and evaporator (Fig. 12.1). The refrigerant is circulated in these components. The performance of these components is very important to achieve overall efficiency of the system. The selection of appropriate matching size of each component is very important for smooth and efficient working of the refrigeration plant. In addition to these major components, many accessories and safety controls are incorporated in the refrigeration system for improving the COP and safety of the plant. The various components commonly used in refrigeration plant are discussed in brief in the following section.

### 12.5.1   COMPRESSOR

The compressor is the heart of the vapor compression refrigeration system. The function of compressor is to suck the refrigerant gas from the evaporator and to compress it to discharge pressure, so that condensation of gas can be done either by water or air at ordinary room temperature. The compressor also keeps low pressure in the evaporator for efficient evaporation of the refrigerant. There are three main groups of compressors:[13] reciprocating compressor, rotary compressor, and centrifugal compressor.

The reciprocating compressor consists of a piston moving back and forth in the cylinder with suction and discharge valve arranged to allow pumping to take place. The rotary and centrifugal compressors have rotating members but the rotary compressor has a positive displacement, whereas the centrifugal compressor draws the vapor and discharges it at high pressure by centrifugal force.[13] Reciprocating and screw compressors are widely used in dairy and food plants. The cooling of compressor head is mostly done by jacketing the cylinder wall and head, through which water is being circulated in large capacity ammonia plants. In case of R-134a or R-22 refrigerant, the discharge temperature will be lower than ammonia compressor and hence, generally fins are provided which facilitates the transfer of heat to the surrounding air.

## 12.5.2 CONDENSER

Condenser is a heat exchanger in which heat transfer[7,8] from refrigerant to a cooling medium takes place. The heat from the system is rejected either to the atmosphere or to the water used as cooling medium.[17,25] In evaporative condenser, water and air both are required to cool the refrigerant. The water is used as cooling medium which in term rejects the heat to the atmosphere. The selection of condenser mainly depends on the capacity of refrigerating system.[5,16,28] The amount of heat rejected or transferred by the condenser is termed as condenser load or condenser capacity.

Condenser capacity (kJ/s) = (Mass flow rate of refrigerant, kg/s) ×
(Enthalpy change of refrigerant while passing
through the condenser, kJ/kg) (12.2)

The condenser load per unit of refrigeration capacity is known as heat rejection factor (HRF):

$$HRF = \frac{Q_c}{R_e} = \frac{R_e + W}{R_e} = 1 + \frac{W}{R_e} = 1 + \frac{1}{COP},$$ (12.3)

where $Q_c$ = condenser load; $R_e$ = refrigerating capacity; and $W$ = work of compression.

Thus, it is clear that HRF depends on the COP of the refrigeration system. If COP is higher, HRF will be lower.

There are three types of condensers commonly used in VCR system; namely, air cooled condenser, water cooled condenser, and evaporative condenser.[28] The atmospheric air is used as a medium of heat transfer in air cooled condensers, while water is used as cooling fluid in water cooled condensers.[8,25] Cooling tower is required for cooling of water used in water cooled condensers. The evaporative condenser uses both water and air for the condensation of the refrigerant.[28] The evaporative condenser consists of a coil in which the refrigerant is flowing and condensing inside and its outer surface is wetted with water and is exposed to a stream of air to which heat is rejected principally by evaporation of water. The water is spayed over the pipes carrying hot refrigerant vapor and movement of air over the wet tubes creates evaporation of water resulting in cooling of the refrigerant. The heat lost from the refrigerant is carried by the air–water mixture leaving the condenser. The heat transfer performance of condenser is very important to get lower condensing pressure.[8,25]

### 12.5.3   RECEIVER

Receiver of the refrigeration plant is a storage vessel of refrigerant of the system and it has to play a very important role to evacuate the part of the system for maintenance and repair. The size of the receiver should be such that it can hold the entire charge of the refrigerant with 1/4 volume available for expansion and safety. It is provided with pressure gauge, purging valve, level indicator, etc.

### 12.5.4   EXPANSION VALVES

The function of expansion device is to reduce the pressure of the refrigerant coming from the condenser as per the requirement of the system and regulates flow of the refrigerant as per the load on the evaporator. There are different types of expansion valves namely: hand expansion valve, high side float valve, automatic expansion valve, low side float valve, thermostatic expansion valve, capillary tube, etc., and they are used depending on the capacity of the system and type of evaporator.

### 12.5.5   EVAPORATORS

The evaporator is a heat exchanger where the actual cooling effect is produced. The evaporator receives the low pressure refrigerant from the expansion valve and brings the material to be cooled in contact with the surface of the evaporator. The refrigerant absorbs the heat from the materials to be cooled (air/water/milk/any other material). The refrigerant takes up its latent heat from the load and becomes vapor. The refrigerant vapor produced in the evaporator is pumped by the compressor and low pressure is maintained to achieve low evaporating temperature. The selection of material to be used for construction of evaporator is based on several factors such as type of refrigerant, thermal conductivity of the metal, cost, ease of fabrication, product to be cooled, etc. Copper is commonly used in small capacity plants due to its higher conductivity and ease of fabrication, but it cannot be used in ammonia plant as ammonia is corrosive to copper. Steel tubes/ pipelines are commonly used in large capacity ammonia plants. The selection of metal for fabrication is mainly decided by the product to be cooled. The evaporator is fabricated from stainless S. S. for icecream freezer, milk cooling equipment, etc., while evaporator of ice-bank system is made from steel tubes. Air cooling evaporators for cold rooms, blast freezers, air-conditioning, etc., have finned pipe coils with fans to blow air over the coil.[26,27]

## 12.6   BALANCING OF DIFFERENT COMPONENTS OF THE SYSTEM

Balancing of various components of the refrigeration plant is a very important design aspect of the system. Misbalancing of any component may greatly affect the performance of the system. Under size expansion valve may result in lower evaporating pressure and over size expansion valve supplies higher flow rate of refrigerant leading to liquid pumping of refrigerant. The rate of the heat transfer plays a very important role in the design of evaporator and condenser.[8,25] It is necessary to pump the vapor from the evaporator at the rate it is produced in the evaporator. If the compressor is not pumping the vapor produced in the evaporator, then pressure of the evaporator increases and it will not be possible to achieve the desirable temperature of cold storage. It is also obvious that evaporation of refrigerant in the evaporator takes place, depending on the load, and therefore capacity control of compressor is one of the essential requirements in economical working of refrigeration plant.

## 12.7   REFRIGERANTS FOR THE REFRIGERATION SYSTEM[4,6,12,14]

The refrigerant is a medium of heat transfer in VCR system which absorbs the heat from low temperature region (i.e., evaporator) and rejects the heat at higher temperature region (i.e., condenser). Any medium which absorbs heat through evaporation is known as refrigerant. Thus, it is a medium of heat transfer which absorbs heat by evaporation at low temperature and gives up heat by condensing at high temperature and pressure. The working fluid in a refrigerating system that absorbs heat from the place to be cooled or refrigerated can be termed as a refrigerant. This heat transfer generally takes place through a phase change of the refrigerant. The refrigerant is a heat carrying medium which during its cycle (i.e., compression, condensation, expansion, and evaporation) in the refrigeration system absorbs heat from a low temperature system and discards the heat absorbed to a higher temperature system. Primary refrigerants are used in VCR system as heat transfer medium which passes through the cycle of evaporation, compression, condensation, and expansion. Secondary cooling mediums (such as water, antifreeze solutions, etc.) are used for cooling. Cooling mediums are first cooled by the primary refrigerant and are used for cooling of air and other products.

Chloro-fluoro carbon (CFC) and hydro chloro-fluoro carbon (HCFC) refrigerants are harmful to the environment. These refrigerants will be phased out universally to protect the environment. Ammonia is a widely used refrigerant in India for large capacity refrigeration systems in dairy[22] and food plants, as it gives a very high refrigerating effect. The thermodynamic properties of some refrigerants are given in Table 12.1.

**TABLE 12.1**   Thermodynamic Properties of Selected Refrigerants.

| Refrigerants | Ammonia (R-717) | Monochloro-difluoro-methane (R-22) | Tetrafluoro-ethane (R-134a) |
|---|---|---|---|
| 1. Boiling point at 760 mm of Hg | −33.3 | −41 | −26.15 |
| 2. Compression ratio at −15°C, +30°C | 4.94 | 4.05 | 6.72 |
| 3. Condensing pressure at 30°C, bar | 11.67 | 12.03 | 7.70 |
| 4. Critical pressure, bar | 113.86 | 49.38 | 40.56 |
| 5. Critical temperature, °C | 133 | 96.0 | 101.1 |
| 6. Evaporating pressure at −15°C, bar | 2.36 | 2.97 | 1.64 |
| 7. Freezing point, °C | −77.8 | −160 | −101 |
| 8. Latent heat of evaporation at −15°C, kJ/kg | 1316.5 | 218.1 | 209.5 |
| 9. Mass flow rate of refrigerant, kg/min/ton | 0.19 | 1.31 | 1.42 |
| 10. Power per TR under standard operating conditions | 0.74 | 0.75 | 0.76 |
| 11. Refrigerating effect at −15°C and + 30°C | 1105.5 | 161.5 | 148.1 |
| 12. Specific volume of refrigerant vapor at −15C, m³/kg | 0.510 | 0.078 | 0.121 |
| 13. Theoretical C.O.P. at −15°C evaporating and + 30°C condensing temperature | 4.76 | 4.66 | 4.62 |
| 14. Vapor displacement per TR at suction temperature of −15°C, m³/min | 0.096 | 0.101 | 0.172 |

## 12.7.1 DESIRABLE PROPERTIES OF REFRIGERANTS

The desirable properties of an ideal refrigerant to be used for VCR system are low boiling and freezing point, high critical pressure and temperature, high latent heat of vaporization, low specific heat of liquid, high specific heat of vapor, low specific volume of vapor, high thermal conductivity, non-corrosive to metal, non-flammable and non-explosive, non-toxic, low cost, easily and regularly available, easy to liquefy at moderate pressure and temperature, ease of leak detection by odor or suitable indicator, high COP, ozone friendly, etc.[15,18] However, there are very few refrigerants which are commonly used as refrigerant. Ammonia is widely used in large capacity refrigeration systems employed for dairy and food plants. Thermodynamic properties of three important primary refrigerants are given in Table 12.1.

## 12.8 DEFROSTING OF EVAPORATOR

The condensation of water vapor of the room/cold storage causes formation of frost over the evaporator. Formation of ice takes place in all the evaporators which are operating below the freeing point of water. The accumulation of ice over the heat transfer surface reduces the heat transfer rate as ice is a poor conductor of heat. Therefore, it is necessary to remove the ice deposited over the evaporator at periodic time intervals. The operation of removing frosted ice from the evaporator is known as defrosting of evaporator. The period of defrosting depends on the type of evaporator, relative humidity of the cold room, evaporation temperature, etc. There are various methods of defrosting such as manual defrosting, automatic periodic defrosting, hot water defrosting, hot gas defrosting, electric defrosting, etc. The selection of the method varies, depending on capacity, type of evaporator, frequency of defrosting required, etc.

## 12.9 PERFORMANCE OF A VAPOR COMPRESSION REFRIGERATION SYSTEM

Coefficient of Performance (COP) is defined as the ratio of refrigerating effect produced to the work of compression. The COP of a VCR plant is mainly affected by factors such as: condensing and evaporating temperature, design of plant components, heat transfer rate at evaporator–condenser, maintenance and service of the plant, multi-stage compression, presence of non-condensable gases in the system, sub-cooling of liquid refrigerant, super heating of suction gas, and volumetric efficiency of compressor.[18]

The effect of evaporating temperature on theoretical COP and the power consumption of refrigeration plant is shown in Table 12.2. The calculated values of theoretical COP and compressor power requirements at various condensing pressures are presented in Table 12.3.

**TABLE 12.2** Effect of Evaporating Temperature on COP and Power Consumption of Refrigeration Plant (Condensing Temperature = 35°C).

| Evaporating Temperature, °C | Theoretical COP | Compressor power kW/ton | % increase in kW/ton |
|---|---|---|---|
| 0 | 6.81 | 0.512 | – |
| −10 | 4.77 | 0.731 | 42.73 |
| −20 | 3.68 | 0.947 | 85.05 |
| −30 | 2.93 | 1.189 | 132.26 |

**TABLE 12.3** Effect of Condensing Temperature on COP and Power Consumption of the Refrigeration Plant (Evaporating Temperature = −15°C).

| Condensing temperature, °C | Theoretical COP | Compressor power kW/ton |
|---|---|---|
| 20 | 6.41 | 0.544 |
| 25 | 5.39 | 0.646 |
| 30 | 4.71 | 0.739 |
| 35 | 4.30 | 0.810 |
| 40 | 3.76 | 0.926 |

## 12.10 ESTIMATION OF REFRIGERATION PLANT CAPACITY

One ton of refrigeration is equivalent to 12,600 kJ/h heat removed by the plant from the evaporator.[9,10] Here, http://ecoursesonline.iasri.res.in/mod/page/view.php?id=3695 indicates following estimations:[9,10]

**Example 1** Assume the bulk cooler capacity = 5000 L; Density of milk = 1.032 g/cm³; Specific heat of milk = 3.9 kJ/kg K; Initial temperature of milk = 35°C; Final cooling temperature = 2°C; Hours to cool = 3.5 h.

Heat to be removed from the milk = 5000 × 1.032 × 3.9
× (35−2) = 664,092 kJ = 6.6 MJ.
Heat to be removed from the milk per hour = 664,092 ÷ 3.5
= 189,741 kJ/h.
The capacity of the refrigeration plant in tons
= 442,723 ÷ 12,600 = 15.

**Example 2** Assume capacity of chilling center =100,000 liters per day;
1 TR = 12,600 kJ/h = 210 kJ/min = 3.5 kW; Operating period of refrigeration plant using ice bank refrigeration system = 18 h; Safety factor
of 20%. Then heat to be removed per day (Q) will be:

$Q$ = 100,000 × 1.032 × 3.89 × (37 − 2) = [100,000 × 1.032
× 3.89 × 35] / [12,600 × 18]
= 62 + [20/100] × 62 = 74.2 TR.

**Example 3** Assume capacity of pasteurizer = 10,000 l/h; Temperature
of milk after regeneration cooling = 15°C. Then heat to be removed
will be:

$Q$ = 10,000 × 1.032 × 3.89 × (15 − 2) kJ/h = 521,882.4 kJ h$^{-1}$
÷ [12,600 TR/ kJ h$^{-1}$] = 41.42 TR.

## 12.11   ADVANCES IN VAPOR COMPRESSION REFRIGERATION SYSTEMS

There are several modifications and advances in the simple VCR system in
order to meet the specific requirements of industry, as well as to improve
the performance in terms of power requirement and maintenance aspects
of the system. These advances includes use of screw compressor, liquid
over feed system, PHE type heat recovery system, ice silo, pre-chillers
for chilled water, design upgradation in different components, capacity
controls and safety controls, cascade system of cooling, etc. Some of these
modifications have resulted in saving of energy.

## 12.12   VAPOR ABSORPTION REFRIGERATION SYSTEMS

The vapor absorption refrigeration system consists of an evaporator, pump,
generator, condenser, and expansion valve. The refrigerant evaporates in

the evaporator by absorbing the heat from space and the vapor is absorbed by a suitable absorbing medium. The solution is pumped to the generator where the refrigerant is expelled from the mixture by application of heat. The refrigerant is then passed into the condenser where it gets liquefied. Heat is the main source of energy for the operation of vapor absorption refrigeration plant. The COP of such systems is around 0.8–0.9, which is relatively very low as compared to a VCR plant and hence, this system is used where heat energy is available at a very low cost.

In ammonia vapor absorption refrigeration system, ammonia acts as refrigerant and water as the absorbent. The ammonia vapor leaving the generator contains some amount of water vapor, which is removed by employing a rectifier.

Lithium bromide-water absorption refrigeration system uses water as refrigerant and LiBr salt solution as absorbent. The LiBr-water system can be used for chilling of water around 4–5°C. In order to perform thermal calculations, enthalpy-concentration diagram of aqua LiBr solution is used. In order to improve the performance of the system, a heat exchanger is incorporated in order to heat the weak solution of LiBr that goes to the generator with the help of weak solution.

## KEYWORDS

- **dairy plant utility**
- **refrigeration plant capacity**
- **glycol cooling system**
- **multi-stage compression performance of refrigeration system**
- **vapor absorption refrigeration system**
- **vapor compression refrigeration system**

## REFERENCES

1. Agrawal, A. B.; Dave, R. K.; Shrivastava, V. Replacing Harmful Refrigerant R22 in Bulk Milk Cooler. *Indian J. Sci. Technol.* **2009,** *2* (9), 974.
2. Arora, C. P. *Refrigeration and Air Conditioning,* 3rd ed.; Tata McGraw-Hill Publication Company Ltd.: New Delhi, India, 2009; p 112.

3. Arora, S. C.; Domkundwar, S. *Course in Refrigeration and Air Conditioning*, 7th ed.; Dhanpat Rai & Co. (P) Ltd.: New Delhi, India, 2006.

4. ASHRAE. Number Designation and Safety Classification of Refrigerants. *ANSI/ ASHRAE Standard;* American Society of Heating, Refrigerating and Air-Conditioning Engineers ASHRAE): Atlanta, Georgia; 1997; p 34.

5. Ayub, Z. H.; John, C. C.; Jon, M. E.; LeCompte, M.; William, R. L.; Robert, P. M.; Manmohan, M.; Ohadi, M. M.; Shriver, G. R. *Condenser.* In *ASHRAE System and Equipment Handbook;* American Society of Heating, Refrigerating and Air-Conditioning Engineers ASHRAE): Atlanta, Georgia; 2000; Chapter 35, pp 1–35.

6. Ballaney, P. L. *Refrigeration and Air Conditioning;* Tata McGraw-Hill Publication Company Ltd.: New Delhi, India, 2009; p 230.

7. Bemisderfer, C. H. *Thermodynamics and Refrigeration Cycles.* In *ASHRAE System and Equipment Handbook;* American Society of Heating, Refrigerating and Air-Conditioning Engineers ASHRAE): Atlanta, Georgia; 2001; Chapter 1; pp 1.1–1.19.

8. Bergles, A. E.; Ohadi, M. M. *Heat Transfer.* In *ASHRAE System and Equipment Handbook;* 2001; American Society of Heating, Refrigerating and Air-Conditioning Engineers ASHRAE): Atlanta, Georgia; Chapter 2; pp 2.1–2.27.

9. Bhadania, A. G. Performance Evaluation and Energy Conservation in Refrigeration and Cold Storages: In: *Compendium: Energy Conservation in Dairy Processing Operations;* SMC College of Dairy Science: Anand, Gujarat, 1998; pp 45–51.

10. Bhadania, A. G.; Patel, S. M. Energy Management in Refrigeration and Cold Storages: In *Compendium: Energy Management and Carbon Trading in Industry;* SMC College of Dairy Science: Anand, Gujarat, 2010; pp 48–52.

11. Camargo, J. R.; Godoy, E.; Ebinuma, C. D. Evaporative and Desiccant Cooling System for Air Conditioning in Humid Climates. *J. Braz. Soc. Mech. Sci. Eng.* **1995,** *27* (3), 1–6.

12. Doerr, R. G. *Refrigerants.* In *ASHRAE Handbook;* American Society of Heating, Refrigerating and Air-Conditioning Engineers ASHRAE): Atlanta, Georgia; 2001; Chapter 19; pp 19.1–19.11.

13. Domingorena, A. A.; Rizzo, D. H.; Roberts, J. H.; Stegmann, R.; Wurm, J. Compressors. Chapter 34. In *ASHRAE System and Equipment Handbook;* American Society of Heating, Refrigerating and Air-Conditioning Engineers ASHRAE): Atlanta, Georgia; 2000; pp 34.1–34.36.

14. Dossat, R. J. *Principle of Refrigeration;* Pearson Education: Noida, India, 2006; 201309.

15. Emmrich, S. K. *Physical Properties of Secondary Coolants (Brines).* Chapter 21. In: *ASHRAE Handbook;* American Society of Heating, Refrigerating and Air-Conditioning Engineers ASHRAE): Atlanta, Georgia; 2001; pp 21.1–21.13.

16. Hajidavallo, E. Application of Evaporative Cooling on the Condenser of Window-Air-Conditioner. *Appl. Therm. Eng.* **2007,** *27,* 1937–1943.

17. Incropera, F. P.; DeWitt, D. P. *Fundamentals of Heat and Mass Transfer.* 4th ed.; John Wiley & Sons: New York, NY; 1996.

18. Jha, S. N.; Aleksha Kudos, S. K. Determination of Physical Properties of Pads for Maximizing Cooling in Evaporative Cooled Store. *J. Agric. Eng.* **2006,** *43* (4), 92–97.

19. Khrumi, R. S.; Gupta, J. K. *Text Book of Refrigeration and Air Conditioning;* Eurasia Publishing House (P). Ltd.: New Delhi, India, 2011.

20. Manohar, P. *Refrigeration and Air Conditioning*; New Age International Publisher Ltd.: Delhi, India, 2010.

21. McQuiston, F. C.; Parker, J. P. *Heating, Ventilating and Air Conditioning-Analysis and Design;* John Wiley & Sons: New York, NY, 1994.

22. Nygaard, H. India's Dairy Sector is Expanding. *Maelkeritidende.* **1996,** *109* (3), 52–53.

23. Parmar, H.; Hindoliya, D. A. Performance of Solid Desiccant-Based Evaporative Cooling System Under the Climatic Zones of India. *Int. J. Low-Carbon Technol.* **2012,** *8* (1), 52–57.

24. Rajput, R. K. *Refrigeration and Air Conditioning.* S. K. Katariya and Sons: New Delhi, India, 2010; 110002.

25. Raol, J. B. *Role of Heat Transfer in Energy Conservation*: In: *Compendium: Energy Management and Carbon Trading in Industry*; SMC College of Dairy Science: Anand, Gujarat, 2010; pp 64–65.

26. Roland, A. A.; Malek, A. H.; Price, G. W. *Air Cooling and Dehumidifying Coils.* In *ASHRAE System and Equipment Handbook*; American Society of Heating, Refrigerating and Air-Conditioning Engineers ASHRAE): Atlanta, Georgia; 2000; Chapter 21; pp 21.1–21.16.

27. Trott, A. R.; Welch, T. C. *Refrigeration and Air-Conditioning*; Butterworth Heinemann: Oxford, Boston, MA, 2000.

28. Vrachopoulos, M. G.; Filios, A. E.; Kotsiovelos, G. T.; Kravaritis, E. D. Incorporated Evaporative Condenser. *Appl. Therm. Eng.* **2005,** *25,* 823–828.

29. Wang, S. K. *Handbook of Air Conditioning and Refrigeration.* McGraw-Hill: New Delhi, India, 2001.

30. Wang, S. K.; Lavan, Z. Air-Conditioning and Refrigeration. In *Mechanical Engineering Handbook*; Frank Kreith, Ed.; CRC Press LLC: Boca Raton, FL, 1999.

## CHAPTER 13

# ENERGY AUDITS AND ITS POTENTIAL AS A TOOL FOR ENERGY CONSERVATION IN THE DAIRY INDUSTRY

A. G. BHADANIA*

*Department of Dairy Engineering, Sheth M. C. College of Dairy Science, Anand Agricultural University, Anand 388110, India*

*\*E-mail: bhadania@gmail.com*

## CONTENTS

## ABSTRACT

Dairy industry is engaged in the processing of milk and manufacture of various dairy products, which require electrical and thermal energy for various dairy processing operations. Energy management is very important, not only for reducing the cost of processing but also to help in the reduction in emission of GHGs in the environment. In this regard, many national, international, and intergovernmental bodies across the world have been working to formulate programs to reduce the energy use in various sectors to combat the phenomena leading to global warming.

Energy audit is the basic requirement for the conservation of energy in dairy plants. Based on the systematic preliminary and secondary energy audits, detailed plan of work for the conservation of energy for the various dairy processing operations can be executed. This chapter covers details on energy audit, principles of energy management, energy analysis of various unit operations involved in dairy plants, and various ways of conservation of energy in dairy plant utilities. The possibilities of energy conservation in lighting, effluent treatment, compressed air supplied, CIP, and all other operations are also discussed with a few case studies. Scope of energy conservation by adopting newer technology, process re-engineering and use of non-conventional energy is also presented.

## 13.1   INTRODUCTION

Energy is one of the important requirements for the growth and progress of any nation. Electrical and thermal energy are required for various dairy processing operations and for providing utilities in dairy plants. The conservation of energy is not only significant for improving profitability of the dairy plants, but also helps in reducing the emission of greenhouse gases (GHGs) in the environment, which are responsible for global warming. Various studies in different countries have shown that significant cost effective energy efficiency improvement opportunities exist in the industrial sector including dairy industry.[20] However, industrial plants are not always aware of energy efficiency improvement potential, unless energy management program is implemented in the industry.

Energy audit is one of the first steps in identifying these potentials. There has been an increasing awareness concerning relationship between economic development, enhanced use of energy and adverse environmental implications due to emission of GHGs. In this regard, many national, international, and intergovernmental bodies across the world have been working to formulate a program to reduce the energy use in various sectors to combat the phenomena leading to global warming. Energy audit is a very essential step in the management of energy for various operations.

This chapter focuses on energy audit and its potential for energy conservation in dairy plants. The purpose of energy audit is to know existing level of energy consumption and to suggest possible measures for the conservation and judicious use of energy. It has become exceptionally critical to manage the use of energy and to adopt all possible measures to conserve it in order to boost the profitability at industrial level and to reduce emission of GHGs.

## 13.2   ENERGY MANAGEMENT

Energy management is a very broad subject covering economical and technical aspects related to the use of thermal as well as electrical energy in various industries.[9,24] However, it is defined as a strategy of adjusting and optimizing energy using systems and measures, so as to reduce energy requirement per unit of output. The basic principles of energy management are:

- Procure all kind of energies at the lowest possible price.
- Utilization of energy at the highest efficiency.
- Reusing and recycling of energy.
- Adoption of most appropriate energy efficient technology.
- Reduce the energy losses.
- Optimization of unit operations and re-engineering process line.

These principles can be successfully implemented by conducting energy audit of each and every operation where energy is used. Therefore, energy audit is one of the requirements for total energy management in order to reduce the cost of energy input to the plant. Energy audit is the key to a systematic approach for decision making in the area of energy management.

Energy management techniques use information generated by energy audit to eliminate waste, reduce and control current level of energy usage and optimization of operating conditions. It is said that "one cannot manage what one does not measure." Therefore, energy audit is the basic requirement for monitoring of energy for various operations. The elements of energy monitoring and targeting are: recording of operating parameters, analyzing information, comparing with standard data, setting targets based on the comparison, monitoring the subsequent processes, reporting new energy consumption data for conclusion, and decision making.

Energy management policy includes: accountability, financial and staffing resources, and reporting procedures and commitment at all levels. It also includes generating awareness, monitoring and reporting system, conducting energy audits regularly, and finally formalizing energy management policy statement. Basically, the four pillars for success of the energy audit are:

- Technical ability,
- Monitoring system,
- Strategic plan,
- Top management support.

## 13.3   ENERGY AUDITS

The conservation of energy is the basic purpose of energy audit in any industry. The information generated by energy audit may help to formulate the benchmark/indices for energy requirement and basis for planning more effective use of energy. It is very essential to provide all utilities in most efficient manner for the conservation of energy. According to Bureau of Energy Efficiency (BEE), energy audit is *"the verification, monitoring, and analysis of the use of energy and submission of technical report containing recommendations for improving energy efficiency with cost-benefit analysis and an action plan to reduce energy consumption."* The objectives of an energy audit can vary from one plant to another. However, an energy audit is usually conducted to understand how energy is used within the plant and to find opportunities for improvement and energy saving.

Energy audit program will help to keep focus on variations, which occur in the energy costs, availability and reliability of supply of energy, decide on appropriate energy mix, identify energy conservation technologies, retrofit for energy conservation equipment, etc.[8] Energy audit reports assist facilities in understanding how the energy is used and to identify the areas for waste reduction and opportunities for improvement. An energy audit is a key for assessing the energy performance of an industrial plant and for developing an energy management program. The typical fundamental steps of an energy audit are:

- Preparation and planning.
- Data collection and review.
- Observation and review of operating practices.
- Data documentation and analysis.
- Reporting of the results and recommendations.

## 13.3.1   TYPES OF ENERGY AUDIT

The type of industrial energy audit to be conducted depends on several considerations such as: Type of industry, purpose of audit, size and type of the industry, the depth to which the audit is needed, and the potential and magnitude of energy savings and cost reduction desired. Based on these criteria, an industrial energy audit can be classified into two types as: preliminary audit (walk-through audit) and detailed audit (diagnostic audit).

### 13.3.1.1   PRELIMINARY AUDIT

In a preliminary energy audit, the available data on various forms of energy used, products manufactured, etc., are collected and analysis of the data is carried out to know the energy performance of the plant. This type of audit is based on the energy used data, which are gathered from electrical bills and thermal energy (coal, gas, or any other) purchased during the last 2–3 years and corresponding production data. It takes less time as it is relatively simple and does not require measurements. A flow chart is prepared, showing the energy flows in the system. These audits provide general basic status of energy use in the plant. Thus, preliminary energy audit provides information on the following aspects:

- Energy consumption status based on existing information and data.
- Analysis of data with respect to month, production, and type of energy.
- Examine the scope for saving energy.

- Identifying the most probable and easiest areas for attention.
- Identify immediate (no cost/low cost) improvement.
- Set a reference point based on the comparison of theoretical data.
- Identifying areas for more detailed energy audit.

## 13.3.1.2   DETAILED ENERGY AUDIT

Detailed energy audits require measurement of energy input for various processing operations. When systems like milk evaporators and dryers are complex, requiring electrical thermal energy at different components of the system, it is necessary to draw the block diagram of the system to understand the energy inputs in the system. The main limitation of energy measurement is the lack of measuring instruments, especially for steam, for various operations. The data of energy consumption and related information collected from all the equipments and processes will be used for the energy analysis of the system. It requires the understanding of the process, energy input, measurement, analysis of data, etc., which takes more time as compared to preliminary audit. The conclusion derived from this audit is more reliable and gives much better energy performance of the system. This audit provides a base for economic analysis of the process and payback period of newer technology, having potential for energy conservation. Following are the three phases to carry out in detailed energy audit:

1. Pre-audit activities.
2. Detailed audit activities.
3. Post audit implementation and follow up.

Pre-audit phase activities include resource planning, finalizing energy audit team, data collection, organizing meeting with various departments of the plant, fixing the time limit, finalizing plant activities, etc.

## 13.3.1.3   SELECTION OF ENERGY AUDIT TEAM

The selection of energy audit team is important for specific industry and the management of the plant should take the decision of selecting energy auditing team from the existing staff or to assign the work to an external agency. The responsibilities of the energy auditing team are to prepare for auditing, execution of the work and finally, preparation of the report. It is desirable to select two separate teams for electrical and thermal energy audits considering qualifications and expertise of the team. It is desirable to choose an authorized government approved energy auditing agency. The team should be aware of all the procedural steps such as planning, fixing responsibilities, execution, instrumentation, analysis of data, format for data recording, report preparation, etc. It is also important to get detailed information of demand load of electricity and the reveling power tariff, in order to estimate the potential of power saving.

## 13.4   ENERGY CONSERVATION OPPORTUNITIES IN DAIRY PLANTS

Thermal and electrical energies are required in processing of milk and for the manufacture of various dairy products. Electrical energy is used for the operation of various pumps, motors, fans, blowers, homogenizers, etc. All heating operations are carried out by steam, which is generated by using coal, furnace oil, or PNG. Small capacity dairy units use even wood and agricultural waste; briquette is produced from agricultural biomass as boiler fuel for the generation of steam/hot water. The steam generated in boiler is used in pasteurizers, milk evaporators, air heaters of spray dryers, ghee boilers, CIP cleaning, can washers and all other equipment where heating is required. Large capacity electric motors are required for the operation of refrigeration plant, fans of spray dryers, packaging machines, operation of cream separators, and several pumps for handling of milk, water, and effluents in the dairy industry. There is ample scope for conservation of electrical and thermal energy used in dairy plants by conducting energy audit and adoption of newer technology having potential of energy conservation.

The design and layout of dairy plant has considerable effect on energy conservation. The following points are to be considered in design and layout of the dairy plants:

- Efficient distribution of steam, refrigeration,[5,6,8,9] water, electricity, etc.
- Provision of heat recovery systems.
- Exploiting maximum use of natural illumination.

- Arrangement of equipment.
- Plant location.
- Selection of equipment.
- Pipeline design with minimum energy losses.

The energy audit of all the dairy processing and product manufacturing equipments such as pasteurizer, separator, CIP system, milk evaporators, spray dryers, chilling system, etc., should be carried out in order to get the level of energy input and opportunities for the conservation of energy.

The energy conservation opportunities in utilities and other operations of dairy plant are discussed in the following sections.

## 13.4.1  REFRIGERATION

Vapor compression refrigeration system using ammonia as refrigerant is widely used in India for industrial refrigeration, air conditioning[1,5,10,12,14,16,19,21,23] and cold storages. In dairy and many food processing plants, ice-bank system of refrigeration is used for chilling and processing of milk, while direct expansion air chillers are employed for cold storages. Direct expansion glycol chillers are also being used for chilling of milk in many dairy plants. The refrigeration system uses electrical energy for the operation of compressor and other auxiliary components of the system. It has been found that electricity consumption of refrigeration plant alone is about 50–60% of total electrical consumption of the dairy, depending on the nature of processing operations. The efficiency of refrigeration plant is measured as coefficient of performance (COP) that varies from 2.5 to 4.0 depending on the operating conditions of the system. The important factors affecting the COP are:

- Type of refrigeration system and its components.[5,6,8,9]
- Design aspects of plant components.
- Operating aspects of refrigeration systems.
  - Evaporating pressure (Evaporating temperature).
  - Condensing pressure (Condensing temperature).
  - Sub-cooling of liquid refrigerant.
  - Super heating of suction gas.
  - Heat transfer at evaporator and condenser.[3,13,15,18,22]

- – Presence of non-condensable gases in the system.
- – Volumetric efficiency of compressor.

- Use of multi-stage compression system.
- Preventing maintenance of plant.
- Adoption of energy efficient technology in the system.

   - – Screw compressor.[21,22]
   - – PHE type condensers and pre-chiller.
   - – Liquid overfeed system.
   - – Fan-less cooling towers.[4]
   - – Heat recovery from discharge gas.
   - – Ice silos.

It is essential to select the right type of refrigeration system, considering the temperature requirement and ambient conditions. It is desirable to operate the refrigeration plant under the highest possible evaporating and lowest condensing pressure to obtain better COP of the system.[2] It is also necessary to maintain good heat transfer.[7,15] conditions for effective condensation of the vapor refrigerant. A member of the energy audit team must be an expert in the area of refrigeration to carry out a detailed energy audit, and to understand the need of newer technology to be selected for the system. When compression ratio is more (> 7.0), it is advisable to use two stage compression refrigeration systems to improve COP of the system.[17]

## 13.4.2 ELECTRICITY

It is noticed that about 75–80% of total electricity consumed in dairy plant is by electric motors. The efficiency of the motor varies with load, and at the full load, it is maximum. Therefore, selection of motor for the given application is very vital in the energy conservation point of view. It is suggested to use high efficiency motors in place of old motors. In addition to this, it is desirable not to use repeatedly rewound motors, as rewinding leads to an efficiency loss up to 5%. The electrical tariff for high tension (HT) consumer of electricity is divided into two categories, that is, demand charges and energy charges, so, as to reduce demand charges and line losses within the plant, one should improve the power factor by installing capacitors. Improvement of power factor from 0.85 to 0.96 will reduce 11.5% peak demand and reduce 21.6% losses. Use of variable frequency drive for variable speed applications such as fans, pumps, compressors,[11] etc. avoiding use of oversize/undersize motors further helps in minimizing the energy usage.

### 13.4.3  STEAM

Selection of energy efficient boiler and its components, performance evaluation of boiler at regular intervals and operational management of boiler are important to reduce the heat losses and for conservation of energy. It is necessary to carry out energy audit of boilers and steam distribution systems and to adopt measures for heat losses through flue gas, radiation, excess air, etc.

### 13.4.4  COMPRESSED AIR SUPPLY

Compressed air supply is a relatively less efficient system in dairy plants. Compressed air is required for the operation of pneumatically operated pouch packaging machines and pneumatic control systems. This utility is very energy intensive. The electrical power requirement depends on the type of compressor and other operating conditions of the compressor. It is noticed that the reduction in discharge pressure by 10% saves energy consumption up to 5%. Lower temperature of air at the inlet is desirable, as a decrease in inlet air temperature by 3°C decreases the power consumption by 1%. The use of screw compressor in place of reciprocating one reduces the electricity power consumption.

### 13.4.5  EFFLUENT TREATMENT SYSTEM

Electrical energy is required for the handling of effluents, and the aeration of the effluents during aerobic treatment of the waste water. It is an unavoidable process of the dairy industry to meet the legal requirements for the disposal of waste water. Combination of anaerobic and aerobic digestion systems enhances the efficiency of the effluent treatment plant (ETP). New technology and methods of treatment for waste water should be adopted for generation of renewable energy. The hydro methane reactor and up-flow anaerobic sludge blanket reactor (UASB) processes are anaerobic processes for the effective biodegradation of organic waste into methane. It is found that 0.5–0.75 kWh energy is needed for the removal of 1 kg of chemical oxygen demand (COD) by aerobic process while in anaerobic process; it generates about 1.2 kWh energy from 1 kg of COD removed.

## 13.4.6 SCOPE OF NON-CONVENTIONAL SOURCES OF ENERGY

As the use of non-conventional sources of energy is eco-friendly, there is a scope to use solar energy in dairy plants to conserve energy, as well as to contribute in the carbon footprint cutback. The use of solar water heating system is an established practice for heating of water required for boiler feed water, cleaning applications, etc. As a result of advancement in solar collectors and related technology, it is prudent to use solar energy in dairy plants. Similarly, solar lighting, wind power, biogas, etc., can be considered for the conservation of energy along with preparing CDM projects.

## KEYWORDS

- dairy plants
- effluent treatment
- electricity
- energy audit team
- energy audits
- energy conservation
- energy management
- refrigeration

## REFERENCES

1. Arora, C. P. *Refrigeration and Air Conditioning*, 3rd ed.; Tata McGraw-Hill Pub. Company Ltd: New Delhi, India, 2009; p 112.

2. Avara, A.; Daneshgar, E. Optimum Placement of Condensing Units of Split-type Air Conditioners by Numerical Simulation. *Energ. Build.* **2008,** *40,* 1268–1272.

3. Ayub, Z. H.; John, C. C.; Jon, M. E.; LeCompte, M.; William, R. L.; Robert, P. M.; Manmohan, M.; Ohadi, M. M.; Shriver, G. R. Condenser. Chapter 35; In *ASHRAE System and Equipment Handbook*, American Society of Heating, Refrigerating and Air-Conditioning Engineers: Atlanta, Georgia; 2000; pp 35.1–35.20.

4. Bakaya-Kyahurwa, E. *Energy Efficient Space Cooling a Case for Evaporative Cooling;* Unpublished Report by Mechanical Engineering Dept, University of Botswana: Bag UB0061, Gaborone-Botswana, 2003.

5. Ballaney, P. L. *Refrigeration and Air Conditioning;* Tata McGraw-Hill Publishing Company Ltd: New Delhi, India, 2009; p 230.

6. Bemisderfer, C. H. *Thermodynamics and Refrigeration Cycles.* Chapter 1; In *ASHRAE System and Equipment Handbook;* American Society of Heating, Refrigerating and Air-Conditioning Engineers: Atlanta, Georgia; 2001; p 1.1–1.19.

7. Bergles, A. E.; Ohadi, M. M. Heat transfer. Chapter 2; In *ASHRAE System and Equipment Handbook;* Chapter 2; American Society of Heating, Refrigerating and Air-Conditioning Engineers: Atlanta, Georgia; 2001; pp 2.1–2.27.

8. Bhadania, A. G. Performance Evaluation and Energy Conservation in Refrigeration and Cold Storages. In *Compendium: Energy Conservation in Dairy Processing Operations;* SMC College of Dairy Science: Anand, Gujarat, 1998; pp 45–51.

9. Bhadania, A. G.; Patel, S. M. Energy Management in Refrigeration and Cold Storages. In *Compendium: Energy Management and Carbon Trading in Industry;* SMC College of Dairy Science: Anand, Gujarat, 2010; pp 48–52.

10. Camargo, J. R.; Godoy, E.; Ebinuma, C. D. Evaporative and Desiccant Cooling System for Air Conditioning in Humid Climates. *J. Braz. Soc. Mech. Sci. Eng.* **1995,** *27* (3), 1–16.

11. Domingorena, A. A.; Rizzo, D. H.; Roberts, J. H.; Stegmann, R.; Wurm, J. Compressors. In *ASHRAE System and Equipment Handbook;* Chapter 1; American Society of Heating, Refrigerating and Air-Conditioning Engineers: Atlanta,– Georgia; 2000; pp 34.1–34.36.

12. Dossat, R. J. *Principle of Refrigeration;* Pearson Education: Noida, India, 2006.

13. Hajidavallo, E. Application of Evaporative Cooling on the Condenser of Window-Air Conditioner. *Appl. Therm. Eng.* **2007,** *27,* 1937–1943.

14. Jha, S. N.; Aleksha Kudos, S. K. Determination of Physical Properties of Pads for Maximizing Cooling in Evaporative Cooled Store. *J. Agr. Eng.* **2006,** *43* (4), 92–97.

15. Kachhwaha, S. S.; Prabhakar, S. Heat and Mass Transfer Study in a Direct Evaporative Cooler. *J. Sci. Ind. Res.* **2010,** *69,* 705–710.

16. Khrumi, R. S.; Gupta, J. K. *Text Book of Refrigeration and Air Conditioning;* Eurasia Publishing House (P) Ltd: New Delhi, India, 2011.

17. Manohar, P. *Refrigeration and Air Conditioning;* New Age International Publisher Ltd: New Delhi, India, 2010.

18. Mathur, A. C.; Kaushik, S. C. Energy Saving Through Evaporative Cooled Condenser Air in Conventional Air Conditioning Unit. *Int. J. Ambient Energy.* **1994,** *15* (2), 78–86.

19. McQuiston, F. C.; Parker, J. P. *Heating, Ventilating and Air Conditioning-Analysis and Design;* John Wiley & Sons: New York, 1994.

20. Nygaard, H. India's Dairy Sector is Expanding. *Maelkeritidende*. **1996,** *109* (3), 52–53.
21. Parmar, H.; Hindoliya, D. A. Performance of Solid Desiccant-based Evaporative Cooling System Under the Climatic Zones of India. *Int. J. Low-Carbon Technol.* **2012,** *8* (1), 52–57.
22. Qureshi, B. A.; Syed, M. Z. A Comprehensive Design and Rating Study of Evaporative Coolers and Condensers. Part I. Performance Evaluation. *Int. J. Refrig.* **2006,** *29,* 645–658.
23. Rajput, R. K. *Refrigeration and Air Conditioning;* S. K. Katariya and Sons, Publisher of Engineering and Computer Books: New Delhi, India, 2010.
24. Raol, J. B. Role of Heat Transfer in Energy Conservation. In *Compendium: Energy Management and Carbon Trading in Industry;* SMC College of Dairy Science: Anand, Gujarat, 2010; pp 64–65.

# GLOSSARY OF TECHNICAL TERMS

**Alpha-lactalbumin** is the primary protein in human milk, and is therefore extremely important for infant nutrition. The structure of alpha-lactalbumin is composed of 123 amino acids and 4 disulfide bridges.

**Antibody** (Ab), also known as an immunoglobulin (Ig), is a large, Y-shaped protein produced mainly by plasma cells that is used by the immune system to identify and neutralize pathogens such as bacteria and viruses.

**Batch pasteurization**: In this process, milk is exposed to 62.8°C for a period not less than 30 min.

**Beta-lactoglobulin** is the major whey protein of cow and sheep's milk (~3 g/l), and is also present in many other mammalian species; a notable exception being humans.

**Bovine serum albumin** (also known as BSA or "Fraction V") is a serum albumin protein derived from cows.

**Carbonation** is the process of dissolution of carbon-di-oxide or incorporation of it chemically.

**Cavitation:** Rapid formation and collapse of vapor pockets in a flowing liquid in regions of very low pressure.

**CIP (Clean-In-Place)** is cleaning process in which cleaning the interior surfaces of process equipment, pipes and associated fittings without disassembly.

**Contamination** refers to the introduction or occurrence of a contaminant in food or food environment.

**Continuous process**: In this process, the flow of material or product is continuous. Processing the materials in different equipment produces the products. Each machine operates in a single steady state and performs a specific processing function. Some examples of continuous processes are pasta production, tomato sauce and juice production, ice cream production, mayonnaise production, etc.

**Cow's milk allergy** (CMA) is an allergy triggered by the protein in cows' milk, which affects up to 1 in 20 children under 3 years.

**Creamery** is an intermediary agency where the cream is extracted from the milk. The separated milk and cream/ghee is disposed off to the consumers and Halwais, etc.

**Critical control point** refers to the process step or procedure at which the control can be applied and is essential to prevent or eliminate a food safety hazard or reduce it to an acceptable level.

**Density** is a physical parameter that is used to compare different materials or to study one material under different circumstances. It is a mass per unit volume.

**Dipole action** is caused due to variations in magnetic and electric fields. Also water is key source for interaction with microwave because of its dipole nature.

**Electric conductivity** is the movement of charges traffic in the center of the carrier.

**Enzymatic oxidation** occurs in the absence of air and presence of moisture. It is catalyzed by certain enzymes between water and oil through enzymatic peroxidation, where enzymes found naturally in animal fats (i.e., lipase).

**Epinephrine,** also known as adrenalin or adrenaline, is primarily a medication and a hormone. It is used for a number of conditions including anaphylaxis, cardiac arrest, and superficial bleeding.

**Evaporation** is a type of vaporization of a liquid that occurs from the surface of a liquid into a gaseous phase that is not saturated with the evaporating substance.

**Falooda** (Hindi: फ़ालूदा; Urdu: فالودہ; Bengali: ফালুদা; also Faluda) is a cold dessert and is one of the Indian versions consists of kulfi, translucent wheat-starch noodles and flavored syrup. Some faludas are served as milkshakes.

**Food-borne disease** is an illness resulting from the consumption of contaminated food, or food with pathogenic bacteria, viruses, or parasites and chemical or natural toxins.

**Food handler** refers to any person who directly handles packaged or unpackaged food, food processing equipment and utensils, or food contact surfaces and is therefore expected to comply with food hygiene requirements.

**Food safety** is an assurance that food will not cause harm to the consumer when it is prepared and/or eaten according to its intended.

**Food stabilizer** increases consistency of the finished food.

**Fouling** is process of deposition or accumulation of unwanted materials on inner surfaces of heat exchagers which reduces thermal efficiency and decreases the heat flux.

**Freezing**, or solidification, is a phase transition in which a liquid turns into a solid when its temperature is lowered below its freezing point.

**Ghee** is complex mixture of many fatty acids and glycosides, phospholipids, fat soluble vitamins and carotenoids.

**Good Manufacturing Practices** (GMPs) are the system of practices to ensure that products are produced with consistent quality and adherence to quality standards.

**Gross marketing margin** is the difference between the price paid by the intermediary and that of the price received by the same per unit of milk.

**HACCP** is a system, which identifies, evaluates, and controls hazards which are significant for food safety.

**Halwai** is an intermediary who prepares sweets and sells raw milk and milk products to the consumers. He purchases milk from producers and milk vendors, etc.

**Hazard analysis** is the process of collecting and evaluating information on hazards associated with the food under consideration to decide which are significant and must be addressed in the HACCP plan.

**HTST** (High-Temperature Short-Time) implies that the milk is heated to a required minimum temperature of 161°F for 15 seconds. Milk is pasteurized to kill any pathogenic bacteria that may be present.

**HTST Pasteurization:** HTST stands for High Temperature Short Time. In this process milk is heated to a temperature of 71°C for 15 s.

**Homogenization** is any of several processes used to make a mixture of two mutually non-soluble liquids the same throughout.

**Hurdle technology** involves elimination or control of pathogens in food products combining more than one approach.

**Hygiene** refers to the conditions or practices conducive to maintaining health and preventing disease, especially through cleanliness.

**Immunoglobin** refers to any class of proteins present in the serum and cells of the immune system, which function as antibodies.

**Immunoglobulin-E (IgE)** are antibodies produced by the immune system. These antibodies travel to cells that release chemicals, causing an allergic reaction.

**kilodalton** is an atomic mass unit equal to 1000 Daltons; usually used to describe the molecular weight of large molecules such as proteins.

**Kulfi** is a popular frozen dairy dessert from the Indian Subcontinent. It is often described as "traditional Indian ice cream."

**Lactoferrin** is a protein present in milk and other secretions, with bactericidal and iron-binding properties.

**Low Temperature Long Time (LTLT)** involves heating the milk to 62.5°C/144.5°F and holding this for 30 minutes, and is the method used by milk banks which perform either the Holder method of pasteurization or the similar Vat method of pasteurization.

**LTLT Pasteurization:** Low Temperature Long Time process in which milk is heated to temperature of 63°C for 30 min.

**Marketable surplus** refers to the total quantity of milk left with the producer after meeting the normal requirements of his family.

**Marketing channels** refer to the sequence of various agencies through which milk and milk products pass on from the producer to the ultimate consumers.

**Marketing costs** represent the expenses incurred by the different agencies involved in the marketing of the milk and milk products. Marketing costs include the cost of milk collection, rent of shop of the intermediary along with dairy utensils and their maintenance, cost of transportation and distribution of milk etc.

**Microwave pasteurization** is done by exposing the milk to microwave energy in batches to have a temperature of 72°C for 15 s.

**Milk pasteurization** is thermal treatment to each milk molecule at a temperature less than 100°C.

**Milk producers** are individuals who own milch animals and supply milk either to the consumers directly or to various milk marketing agencies viz., milk vendor, Halwai, cooperative societies/collection center, creameries, etc.

**Milk vendor** is a functionary who collects milk from the producer and sell it either direct to the consumer or to other agencies.

**Net marketing margin:** When from the gross marketing margin of an agency the cost borne by it are deducted, the remainder is called the net marketing margin or profit margin.

**Nusselt number:** In heat transfer at a boundary (surface) within a fluid, the Nusselt number (Nu) is the ratio of convective to conductive heat transfer across (normal to) the boundary. In this context, convection includes both advection and diffusion. It is named after Wilhelm Nusselt, it is a dimensionless number.

**Ohm's law** states that the current flowing in the resistance is directly proportional to the applied voltage and inversely with resistance.

**Ohmic heating technique** has an important role in elimination of live microbes' presence in milk.

**Organoleptic properties** are the characteristics of food products, water and other subtances that an individual experiences through the senses such as taste, smell, touch, and sight.

**Pasteurization:** To expose a food to an elevated temperature for a certain duration sufficient to destroy microorganisms, that can cause disease or cause spoilage or undesirable fermentation of food, without radically changing organoleptic quality.

**Pathogen** is an organism which has ability to cause cellular damage by establishing in the tissue, leading to clinical signs with result of morbidity or mortality.

**Phosphatase** is an enzyme that removes a phosphate group from its substrate by hydrolyzing phosphoric acid monoesters into a phosphate ion and a molecule with a free hydroxyl group.

**Prerequisite programs** refer the practices or procedures followed in manufacturing facility to control the hazards, thereby preventing the contamination of the food product.

**Price spread** is the difference between the price paid by the consumer and that received by the producer for an equivalent quantity of the product.

**Psychrophilic:** These are capable of growing and reproduce in cold temperatures, ranging from −20 °C to +10 °C.

**Sanitation** refers to the creation and maintenance of conditions favorable to good health.

**Sonication** is the act of applying sound energy to agitate particles in a sample, for various purposes.

**Specific heat** is the amount of energy to raise water temperature by one degree at a constant pressure.

**Spray drying** is the transformation of feed from a fluid state into a dried particulate form by spraying the feed into a hot drying medium.

**Sterilization** refers to any process that kills (deactivates) or eliminates all forms of life and biological agents.

**Sweetener** is a food additive that provides a sweet taste like that of sugar while containing significantly less food energy. Some sugar substitutes are produced by nature, and others produced synthetically. Those that are not produced by nature are, in general, called artificial sweeteners.

**Thermal diffusion** represents the thermal conductivity of food divided by its density and specific heat.

**Thermoduric:** Able to survive high temperatures specifically survive pasteurization in case of microorganism.

**UHT Pasteurization** refers to an Ultra High Temperature process in which milk is heated above 130°C temperature required to kill spores in milk for 1–2 s.

**Vacuums pasteurization**: In this method, milk is exposed to different thermal treatments under vacuum.

**Volumetric heating process** is the process in which microwaves penetrate uniformly throughout the volume and delivering heat evenly from center to outer boundaries of product due to dipole action and other associated mechanisms.

# INDEX